MY NAME IS PACOMIO

The Life and Works of Colorado's Sheepherder and Master Artist of Nature's Canvases

*(¡Me Llamo Pacomio!
La Vida y la Obra del Pastor y Artista de Colorado:
Maestro de Lienzos de la Naturaleza)*

BY STEVEN G. BAKER
FOREWORD BY LEIGH ANN HUNT
ILLUSTRATED BY PACOMIO M. CHACON

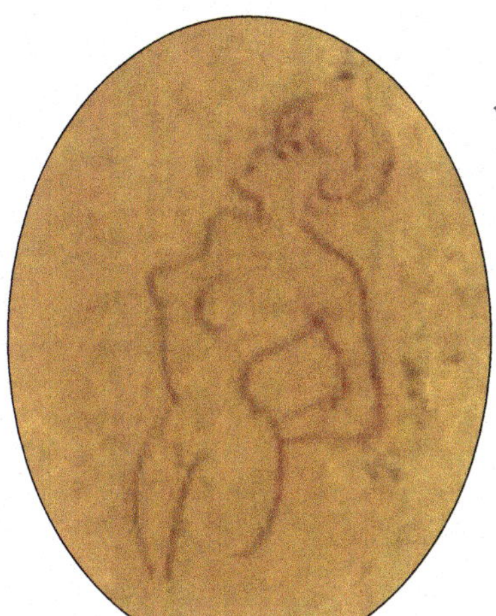

WESTERN REFLECTIONS PUBLISHING COMPANY®
Lake City, CO

Title page image is a classic Pacomio Chacon "pin-up style" nude carved into a metal stock driveway sign in the White River National Forest (image is from Chacon's miscellaneous art portfolio-image PMAP 19 in chapter 9).

ISBN 978-1-937851-14-9
Library of Congress Number: 2016943288

First Edition
Printed in the United States of America

Western Reflections Publishing Company®
P.O. Box 1149, 951 N. Highway 149
Lake City, CO 81235
(970) 944-0110
publisher@westernreflectionspublishing.com
www.westernreflectionspublishing.com

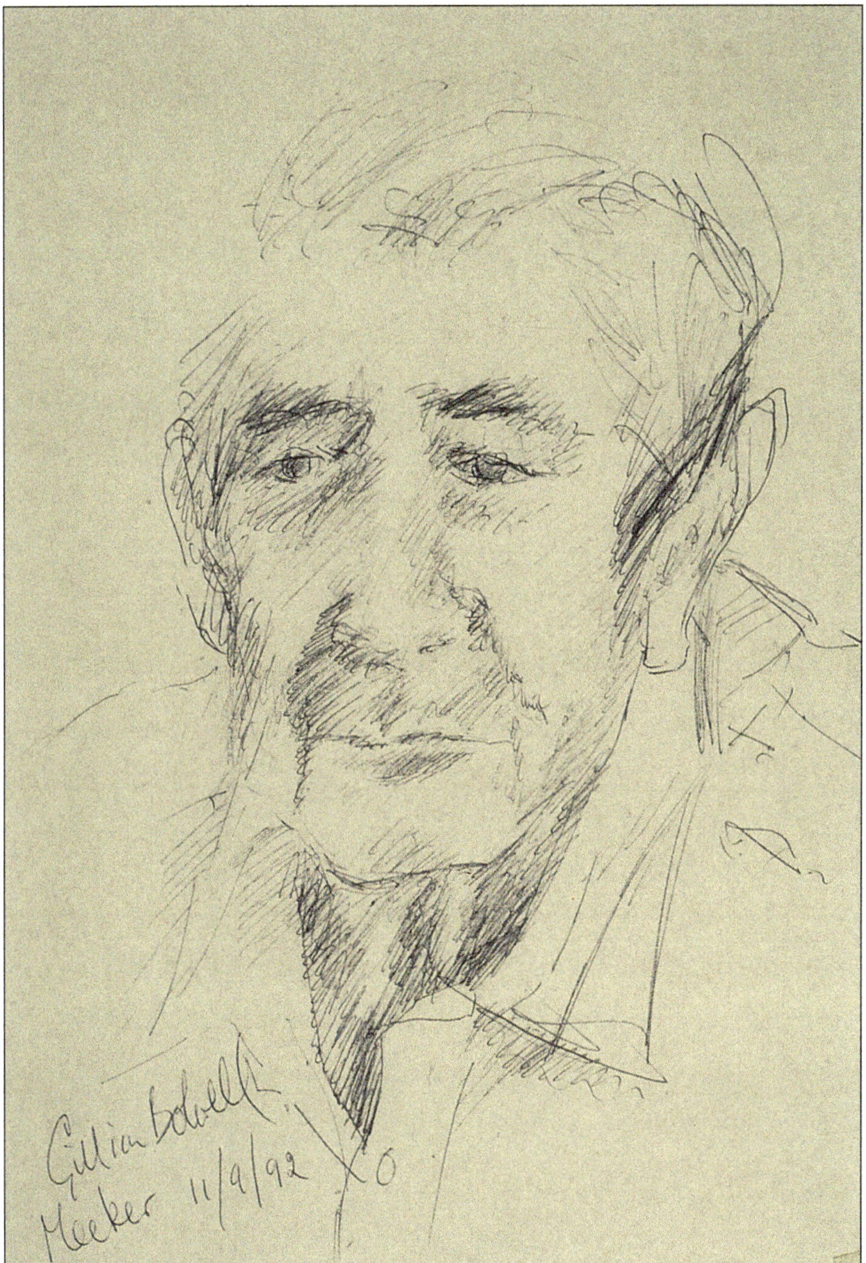

Frontispiece: Portrait of the retired sheepherder and master folk artist, Pacomio M. Chacon (1916-2009) at 76 years old. Drawn by Australian artist, Gillian Bolwell, in September 1992 on a paper dinner place mat at the annual banquet for the Meeker Classic Sheepdog Championship Trials, Meeker, Colorado, where Mr. Chacon was an honored guest (gift to the author from Pacomio Chacon).

Con muchas lágrimas por mi muy buen amigo,
Pacomio Chacon, profesor de las ovejas y
artista maestro de los árboles y los riscos.

de su portavoz y pátron, Esteban

(With many tears for my very good friend,
Pacomio Chacon, professor of the sheep and
master artist of the trees and the cliffs.

from your spokesman and sponsor, Steven)

FOREWORD

I began working at archaeology in the eastern California national forests nearly forty years ago. At that time groves with carved aspen trees were plentiful throughout the forests. The carvings were easily read and widely enjoyed. Whether the subject matter was mundane, factual, poetic, comical, or openly pornographic, the carvings spoke of decades of sheep herders working alone in the woods. We noted the names and dates, mostly Spanish Basque in that region, and mostly 1900-1940s, and moved on. We did not consider the carved trees to be very significant as cultural resources, since they were then so common, and similar herders continued to herd sheep in those same places. Since then, having worked in forests in Nevada, Utah, and western Colorado, I have seen and documented thousands more aspen carvings and have recently spent time relocating examples that were recorded in the earlier years of surveys in the Uncompahgre National Forest.

But lately, amid the immense aspen forests of western Colorado, the carved trees have been getting harder to find and even harder to read. They are dying off and the surviving carved letters and images are blurring and running together. By now many people have forgotten that sheep once were herded just about everywhere. Most of the ranches in local towns had flocks up in the mountains or on the deserts.

Now we look on the remaining aspen carvings with new eyes, weighing the significance of the notations and artwork of generations of anonymous sheepherders. Archaeologists have begun to consider sheepherder carvings to be representative of a whole pattern of economic interaction, the only trace left of the men and lads who left their homes to work for distant ranches, earn some cash, and return home whenever they could. Databases are being built with the names and home towns (often included by the carvers) and dates, to connect the family lines of herders, sheep grazing permittees, economic trends, regions of employment, and issues of range management. The carvings also provide insights into the daily lives, concerns, and interests of the men. We ponder and romanticize the lonely, high-altitude life of sheepherders and we have come to appreciate the trees so much more, even as the carvings they bear are fading away.

Carving aspens as the Hispanic and Basque herders traditionally did it is completely distinct from other kinds of graffiti, such as the hacking of initials and hearts on the trunks. It requires patience and repeated site visits to see the results. The delicate cuts made through the white paper bark into the pale green layer below are barely visible until the cuts have healed into grey scars. A carver hardly knows what the finished image will look like unless he is able to come back months or years later to inspect his work.

One herder's work from eastern Utah and the Western Slope of Colorado obviously stands head and shoulders above the rest. Pacomio Chacon had a truly impressive way with carvings.

His figures, and even his earliest signatures, had an artistic way with line weight, shading, and fluidity. Locally his works became famous. Surrounded by thousands of blank aspen canvases in the summer ranges and inviting rocky cliffs of the winter ranges, he developed his art to a high level. His life was typical of so many of those from his time and place, many of whom I have personally known. Chacon's origins, life story, and values, were common among those employed among Colorado's mountain flocks. Through his art he highlighted and symbolized for us the lives of so many herders who will never again watch and worry over flocks throughout the West.

Steve Baker's book captures and preserves Chacon's biography and artistic legacy with sensitivity and affectionate detail, and displays many examples that are extinct today. Pacomio Chacon's carvings were a gift to us, and this book will let them endure for perpetuity in the annals of Western history.

Leigh Ann Hunt
U.S. Forest Service (Retired)
Eckert, Colorado
December 2015

AUTHOR'S PREFACE

It is my great pleasure to introduce my readers to the Colorado sheepherder and aspen and cliff carving "master" folk artist, Pacomio ("Paco") Martinez Chacon (1916-2009), and his artwork. Paco's now rapidly disappearing art can only be viewed in a few private collections or in remote rural areas of western Colorado and eastern Utah. Accordingly, up to now it has not yet been widely recognized. This book--the first written about him--is now on the cusp of making Paco's exceptional work known to a wider public.

As early as 1992, Paco had already been designated a "master" folk artist by the Colorado Council for the Arts, the Museum of Western Colorado, and the Rocky Mountain Regional Office of the U.S. Forest Service and some of its National Forest offices. During Paco's lifetime these agencies and most of the individuals who were even remotely aware of his work, informally considered him a living Colorado treasure and probably the finest carver of aspen trees and rocky cliffs who ever walked the old sheep ranges of the American West. This book documents Paco's portfolio of ephemeral artwork and tells his life story within the context of sheepherding. The volume is intended to obtain for Paco and his work the recognition it deserves within the context of Colorado and the broader western American folk life and cultural heritage.

It has taken over twenty-five years for me to assemble the pertinent information and write this little book about my friend Paco and his art. A number of people have assisted me through this time. Thanks especially to Phil Born and Monte Sanburg who made up the core of my field crew for the years that I worked in the vicinity of Rangely, Colorado, where we recorded so many of Paco's images in the sandstone cliffs. Phil also once worked for the U.S. Forest Service in western Colorado and recorded some of Paco's aspen carvings for that agency. He was the first archaeologist to interview Paco, and his good notes were a great help in preparing this book. My fellow archaeologist, Jim Truesdale of Laramie, Wyoming, was privileged to work on the Northern Ute Reservation in Utah where Paco had wintered sheep and left cliff carvings in that area. Jim shared his discoveries with me.

Ashton and Ellen Robinson, a professional trapper and photographer duo formerly from Meeker, Colorado, found many of Paco's aspen carvings in the White River National Forest near there, documented them, and shared them with me. I am grateful to them and fondly recall the time I spent in their company.

Thanks also to Dave Cogswell of Meeker for allowing me to photograph his collection of Paco's aspen carvings, as did the late Dick Moyer of Meeker. Dick was a logger and outfitter who discovered and salvaged a great many of Paco's aspen carvings over the years. Thanks as well to his son, Dave Moyer, and the rest of the Moyer Family for tentatively agreeing to donate examples to the Museum of Western Colorado. Unfortunately, the last word I received as this volume went to press was that all of the surviving examples of the Moyer collection were either given away by his father in his later years or were stolen from his barn in Meeker. It is, therefore, doubtful that any of Paco's artwork will ever be curated in a museum collection. What Mother Nature has not already obliterated has now been lost. There will be few opportunities in the future for anyone to see it except in this book.

Leigh Ann Hunt, former lead archaeologist for the Grand Mesa, Uncompahgre, and Gunnison National Forests in Colorado, has long been a serious student of sheepherding culture. She provided me with helpful information and examples of Paco's work. I would like to express my gratitude to her for her ongoing encouragement, collaboration, and preparation of the foreword for this book. I am also grateful to the Colorado Council for the Arts for providing a grant that allowed me to take Paco to the Meeker Classic Sheepdog Championship Trails in 1992 and to prepare a formal exhibit of his artwork shown there and in other locations. This grant allowed me to begin working closely with Paco and start making this record of his life and work. Ronna Lee Sharpe and Mike Perry of the Museum of Western Colorado were instrumental in obtaining the grant from the Arts Council on Paco's and my behalf. We both appreciated their efforts.

Aspen art researcher, Peggy Bergon of Pagosa Springs, Colorado, shared examples of sheepherders' art with me. Ernie Etchart of Etchart Ranches in Montrose, Colorado, read portions of the manuscript and suggested some helpful corrections that addressed some aspects of sheepherding from the perspective of the sheep ranchers who employed the herders. In 2014 Ernie arranged for me to photograph one of his flocks while it was being trailed to its spring grazing grounds near Montrose.

For many years Gail Carroll Sargent has been responsible for the professional artwork that accompanies my research completed under the auspices of Centuries Research, Inc. Although now retired, she once again drew upon her considerable artistic abilities in preparing the two, fine, watercolor illustrations included in this volume. Nancy Lamm prepared the map in Figure 2.

Gus Halandras of Meeker, Colorado, was instrumental in establishing the dog trials there and creating the opportunity for us to attend and for Paco to conduct a workshop on aspen carving. Paco and I appreciated his efforts on our behalf. He also obtained images of some of Paco's work for us. Many of Paco's rock carvings were recorded with the sponsorship of my former client, Chandler and Associates of Denver, and their Southwest Rangely Project in Rio Blanco County. Staff, including Troy Osborn and Patrick Uphus from the Meeker office of the White River National Forest, also shared examples of Paco's aspen carvings. John Ogden of Moab, Utah, provided photos of Paco's lovely "Canyon Bride" (image PRA 4 in Chapter 8) from Salt Wash near Arches National Park, Utah. On my own and Paco's behalf I thank everyone for their contributions.

I must also thank my son, Matthew Baker, for helping to locate and photograph images of Paco's work from Utah. Thanks also to my wife, Nancy Ellen, and our family friend, Dorothy Causey, for proof reading and otherwise constructively commenting on this offering. I have to admit that I was not fully attentive over the years as people provided me photos of Paco's art. For this reason, along with a massive computer failure that destroyed many e-mail records, I have not always been able to identify the sources of some of my images. I apologize for any resulting oversights in recognizing individuals who provided examples of Paco's work or otherwise contributed to this project.

Thanks also to Connie Theos of Meeker; Mike Theos, Jr. of Denver; and Toula Theos of Cody, Wyoming; who provided information covering the years that Paco worked for Toula's late former husband, Mike Theos. Gillian Bolwell is an Australian artist who won the poster design competition for the 1992 Meeker sheep dog trials. Thanks to Gillian for sketching Paco's likeness on a paper dinner place mat at that year's banquet and giving the image to Paco and me for use as the frontispiece. Jim Houston, formerly of Gunnison, Colorado, also made some of Paco's artwork available. My good and ever reliable colleague, Rick Hendricks, New Mexico's state historian, voluntarily edited this volume and assisted me with some of the New Mexico history and the Spanish language and customs. Thanks once again Rick!

Paco's brother, Presciliano, (1923-2013 [aka "Press"]) was often present when I spent time with Paco and contributed to my efforts to assemble the family history. Press was also a sheepherder who occasionally carved both in the aspens and on the cliffs. Although his distinctive works also reflect some emergent talent, he was not as accomplished or prolific as Paco. Because Press's work can easily be confused with Paco's, some further information about his work is provided in the discussions that accompany Paco's artwork in Chapter 6 and in his portfolios.

Paco; his daughters, Alice Montano, Susanna Tipia, Sofia Chacon, Marcella Aragon, Priscilla Studt; and his son, Pacomio Chacon, all encouraged this book. Alice, in particular, took the lead in helping to gather family source materials and information. I could not have completed this project without her interest and assistance. Very little historical documentation exists about the Chacon Family. Most of the narrative of Paco's life included herein has, therefore, been drawn from family members or interviews my associates and I conducted over many years. I am grateful for Alice's help and the encouragement of all of Paco's family.

When Paco finally crossed over the Great Divide in 2009, his family and friends embraced me as part of the family and accorded me the honor of contributing to his eulogy at his funeral. They also encouraged me to tell Paco's story "straight" and not to feel obligated to "sanitize" his human failings while emphasizing his personal strengths and skills.

In March 2013 my wife, Nancy Ellen, and I took a brief vacation to Las Vegas, Nevada. While there we visited the showroom of the noted sculpter, Richard MacDonald, who has completed outstanding metal sculptures of performers in the Cirque du Soleil. A great many of these are absolutely stunning females. One simply has to see these exquisite pieces to understand the phenomenal artistic ability of their creator. As I toured the showroom I could only think how much Paco would have enjoyed seeing these pieces of art. I am confident that he would have been rendered speechless beyond perhaps muttering a few low "ay yay yays" to show his appreciation.

I do not know whether Paco ever visited an art gallery other than those displayed in magazines. I would have loved to have taken him to some and shown him the works of the grand masters. Seeing MacDonald's work made me think of Paco, and it was there, surrounded by all of these beautiful feminine sculptures, that I decided that it was finally time to write Paco's book, even if I had not fully completed my plan to gather as much of his work as possible. Like Paco, I, too, was turning greyer and running out of time faster and faster. Simply put, it was now or likely never! I thus completed most of this volume, "Paco's libro," in the late spring of 2013.

Finally, thank you so very much Paco for your loyal friendship, companionship, and teachings. I can only hope I did you justice. Words cannot describe how terribly sorry I am that I could not complete your libro while you were still among us and could hold it in your hands and read it, take pride in it, and be convinced that your artwork truly counted and has been preserved for posterity. I look forward to one day reconnecting with you on the other side of the Great Divide. I do not know if there are aspens there for you to carve on in "sheepherder heaven," but I am confident that the Lord will provide lambs for you to tend to and to teach me about. So, I know you won't be idle until I can also get there! I am proud to have known you and am honored to tell the world that you truly were "PACOMIO: MASTER OF MOTHER NATURE'S CANVASES!"

Steven G. Baker, Patrón
Montrose, Colorado
December 2015

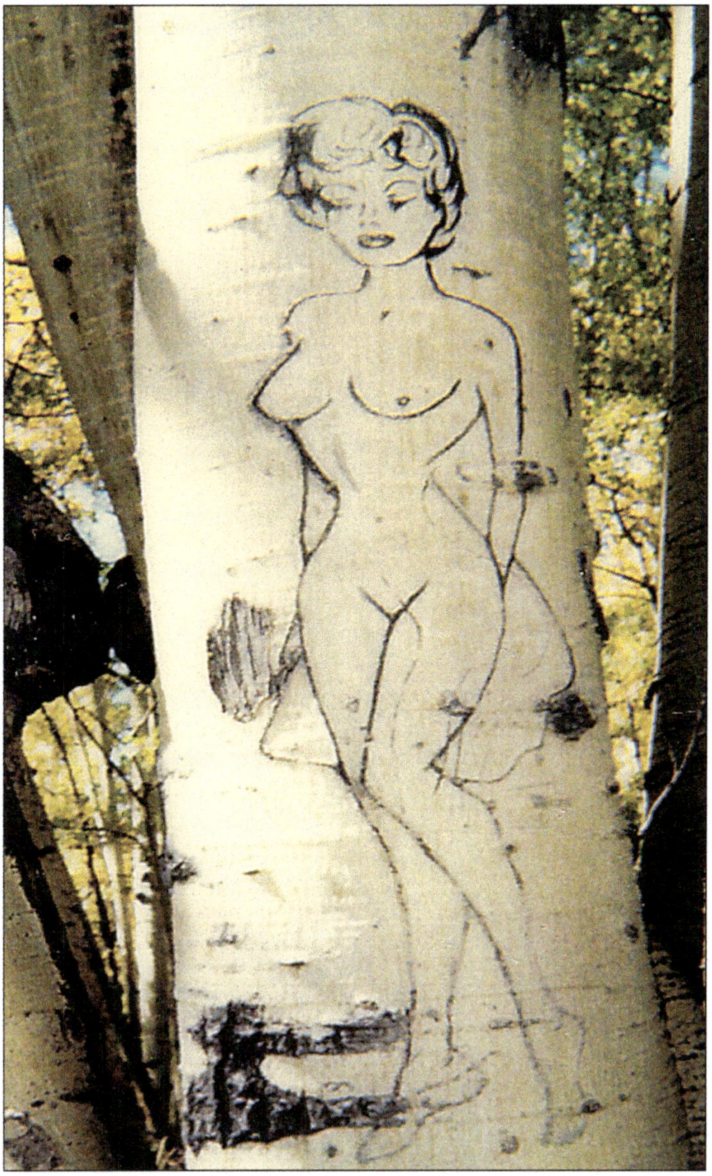

Figure 1: Paco's undated (probably 1970s) "Marilyn M.-Chimera of the Aspens," one of the artist's finest works and classic "calling cards" at the height of her beauty, from the White River National Forest, Meeker, Colorado (Paco's aspen portfolio image PAP-19).

TABLE OF CONTENTS

LIST OF FIGURES

LIST OF TABLES

Chapter 1

AUTHOR'S INTRODUCTION

The Bible's Book of Revelations explains how time is a thief that erases life's accomplishments. True to that notion, time is certainly doing this to the extensive but very impermanent art portfolio of the little sheepherder, Pacomio Martinez Chacon (1916-2009), of Fruita, Colorado. "Paco," as he was generally known, was a self-taught folk artist who, for roughly sixty years, left his prolific and very distinctive signature and carvings out on the aspen trees and rocky cliffs of the remote sheep ranges of western Colorado and eastern Utah. Until now Paco's art could only be seen by those fortunate few (such as sheepherders, hunters, archaeologists, loggers, foresters and the like) who had reason to routinely be out in the woods and canyons and were lucky enough to chance upon it.

Among this small but growing group of admirers, Paco became something of an icon, particularly while his rapidly disappearing aspen carvings still flourished. They came to know him as one of the most talented, if not the most talented, of all the sheepherder artists who ever worked the ranges of the American West. As my readers will note as they peruse his artwork, this is not an idle boast made with the aim of bragging on a friend.

Although sheepherders very commonly carved pictures, their names, and varied testaments in the aspens, those who know Paco's work challenge anyone to find one who did so with such a consistently high artistic quality as he did! If such an individual existed, I and my colleagues certainly would like to hear about it! Extensive research and inquiry throughout the Western U.S. among those who record such images have failed to reveal any other artists who even came close to competing with Paco's artistic ability.

Since simple serendipity brought me into personal contact with Paco more than twenty-five years ago, I have been working to document as much of his artwork and personal history as I could. In this effort I have been most fortunate. With the assistance of Paco, his family, and a few of his other friends and admirers, I have finally managed to accumulate enough information to produce this little tribute to him and his work.

I have written of many people and things during the course of my career as a historian, anthropologist, and archaeologist, which spans more than a half-century. When you write of the dead, especially personal friends, there is always an extra layer of responsibility to portray their lives sensitively, accurately, and respectfully.

They are no longer around to correct you, defend their reputations, or push your nose in if you really foul things up. While I trust that I have gotten Paco's story at least reasonably correct, no writing project has ever caused me more personal grief and heartache than this little book. Paco was my good friend and teacher. This easy-going little sheepherder was very proud that someone cared enough to write about him and his work. I grieve for him because he never got to see this book as I promised him for far too long he would. I discuss more about this and other aspects of our close relationship in my personal reminiscences in Chapter 10.

His family and friends certainly mourned him when Paco finally escaped the icy grip of his dementia and passed away on July 21, 2009. Except for their memories and the artwork considered herein, there was little about him that the world would really have known or missed when he passed. He had very little formal education and no wealth. He had no social or political status except among his family, people in the regional sheep industry, and a few of we other folks lucky enough to get to know him and see his artwork on Mother Nature's grand canvases.

Paco spent most of his ninety-three years in the outdoors among the sheep. They were what he knew and understood so very well. He truly loved them, was at home with them, and was a great herder. This fact was recognized and seemingly acknowledged by all the bosses he ever had, even if, when he was on a typical "sheepherder's tear," they had found it necessary to fire him on one or two occasions. When they did, they later regretted it because they lost not only one of their best herders but also the only one who could "doll up" a routine sheepherder's grocery list with beautiful calligraphy and all manner of images, including lovely nude women. He could also very humorously mimic a great variety of people, including all of his former bosses. He was quite a mime and good entertainer.

Although he loved his family, much of Paco's pride and sense of personal self-worth came from his skills as a herder and his God-given artistic ability. Obviously born with a natural talent, which is suspected of running in the family genetics, Paco had absolutely no training as an artist. He was entirely self-taught. He began by emulating examples of art he found in newspapers, catalogs, and particularly men's pin-up calendars and magazines. His ability to recall and portray the human and other life forms from memory accurately without models was outstanding as is evident in his portfolios included herein.

Paco's favorite subjects were women of the "calendar-girl" variety. He admired them and routinely celebrated their natural form in his art. Nude women were his calling cards, and he left them everywhere he could find a surface to draw, scrape, or cut on! Even when he did not sign a drawing or carving, everyone involved in the regional sheep industry or public land management knew that Paco had been there. His images, whether women or animals of the forest, were almost as well recognized as his signature, and they appeared over a wide area.

In this regard there is one thing I wish to very strongly clarify for my readers now, right up front. Although he loved to draw beautiful nude women, Paco was a gentleman! While he was human and certainly not without fault, there was nothing pornographic about most of his art or actions. The only exceptions were a couple of naughtier and remotely located images he admitted to leaving behind on whims. He was nearly always polite in his art and personal dealings.

As a man, and like Paco, I certainly appreciate the natural female form. My focus here is, however, intended to be much more refined than that. Since it is a book about an artist who just happened to have celebrated naked women, as an author I also have to do so. I expect my readers to accord both Paco and me, as the author, appropriate respect relative to this "celebration of women," just as they would respect the works of the "grand masters" whose nudes hang in famous art galleries. My purpose is to celebrate Paco's ability to produce high-quality carvings on the difficult natural mediums of aspen trees where they are referred to as "dendroglyphs" or "arborglyphs" and sandstone cliff faces where they are known as "petroglyphs."

Paco was truly *the* master of these techniques. As previously stated, he intended his art to be his calling card and a bit of a shout to the world that "My name is Pacomio!" It was his way of proclaiming that he was more than just another sheepherder. It was to stake his claim and earn the recognition of his peers and society at large. As a lonely man in the prime of life and living away from his wife, or any other female companionship for extended periods, he often thought about women. But this does not diminish his skills as an artist. If he had only drawn wildlife or landscapes, we would still be celebrating his work!

This little volume is truly Paco's book, his "libro" as he always referred to it. It is intended to preserve examples of his artwork and to introduce and memorialize Paco, his and sheepherding's place in the history of Colorado and the American West, his contribution to ephemeral American folk art, and to what some art scholars would describe as the genre of spontaneous "outsider art" (Rhodes 2000).

Although it is rumored but unproven that noted auction houses such as Christie's have sold some of Paco's aspen art, you cannot go to a gallery and see his work. You have to access this book or stumble upon his images in the wilds by simple dumb luck, just as I first did. A variety of people walking literally thousands of miles out in the remote old sheep ranges over several decades assembled the examples in Paco's portfolios reproduced herein.

Paco's images included in this book were recorded by a number of individuals under varied field conditions with a variety of cameras, different films, digital cameras, and varying lighting. The images are, therefore, not constant in the nature and quality of their recording as found in most "coffee table style" art books where all photography is usually done professionally under ideal circumstances.

In comparison to other kinds of art, the unusual aspect of Paco's aspen tree and rocky cliff-face canvases is that they are so transient and impermanent. The lifespan of his tree art is extremely limited. Once carved it is not readily visible until it has matured for two to three years. It then matures, and the fine scarring from Paco's cuts only leave a good image for a few years. It only temporarily blossoms, much like a flower, before scarring begins to damage and overgrow the images. The images transition from virtually unnoticeable, to full bloom, to heavy scarring in only a decade or two.

Some of Paco's cliff art made on softer sandstone in the 1970s is already well deteriorated. Thus, although some of Paco's rock art may well still be around in another hundred years, the last of his aspen carvings are even now becoming aged. Many have already been lost to fire, decay, or in timber sales offered by the U.S. Forest Service, which for many years did not even record the aspen carvings of sheepherders, let alone attempt to preserve them.

Loggers involved in such sales salvaged some of Paco's work, and these specimens make up a substantial portion of the examples exhibited herein. Others were looted from the forests many years ago, soon after Paco had carved them. Although the trees in a timber sale belonged to the logger who purchased them, people with no legal rights to them took many of Paco's images from the public domain.

It is against this background that I introduce my readers to Pacomio, sheepherder and master artist of Mother Nature's canvases, and give them the opportunity to review and enjoy his artwork. It would be Paco's hope that you, dear reader, will take time to stroll through his portfolio, smile, and chuckle while doing so. If you do so and look closely at "Paco's Ladies," I am confident that you will come to appreciate the individuality and often whimsical and chimerical characterizations that he intentionally imparted to each of them. It obviously takes a lot of talent to carve on a tree a woman who truly looks like Marilyn Monroe, as the artist intended and can be seen in Figure 1. Most people have no problem in recognizing this work as Paco's rendition of Ms. Monroe.

This is one aspect of Paco's works that brought a great deal of joy to me and my assistants as we recorded them, particularly those depicted in his rock art. It was with great pleasure that we worked hard to give them appropriate names by which to remember them and not simply relegate them to sterile archaeological site numbers.

Other folks gathered and/or recorded most of Paco's aspen carvings, but I have taken it upon myself--actually usurped the pleasure--to name many of his ladies of the trees as well as those of the cliffs. He encouraged me to do so but was quite surprised that anyone would bother. Thus, it is my great pleasure to introduce you to Paco's: *Blossom; Marilyn-Chimera of the Aspens; Flouncy; Canyon Bride; Dream in Gossamer; Evening Breeze; Morning Eyes Awaken; Little Miss Muffet; Little Traveler; Ninita Caprichosa* (Capricious Little Gal); and all his other "little gals" carved into the rocks and trees. Visit the *Dueling Artist's Cave*, meet *Kino* (Paco's favorite dog), view *Paco's Nacimieto* (Nativity scene) carved on Christmas Day in 1974, and all the other examples of his work.

Above all, keep a light heart, laugh, and enjoy yourself. If you have any inhibitions about the natural female form, forget them, let them go, and lighten up just as you would at the Denver Museum of Art, the Louvre, or the Metropolitan, all of which also display many nude female images! That admonishment goes for the ladies and men as well. That is all Paco would ask because that is why this man went to such lengths to leave us this truly wonderful testament to his short time on Earth and thereby introduce himself by saying "My name is Pacomio!"

Paco's stated purpose was to leave something behind that someone would discover by accident that would "make them smile and be happy." As you stroll the forests and canyons with Paco, I hope you will also take time to read about him and sheepherding and thereby begin to understand the simple, pastoral, old New Mexico culture that he was born into and the solitary lifestyle he lived for so many decades here in Colorado and Utah. I have also attempted, with a touch of humor and sentimentality, to explain the unusual and strong bond that developed between us, two men from such highly divergent worlds.

The biographical information on Paco and his family was obtained from formal and informal interviews I (Baker 2003a) conducted and from information gathered while Paco and I traveled and otherwise spent time together. Additional interviews with the artist by Phil Born (Born 1987) and Polly Hammer (Hammer 1990) of the U.S. Forest Service provided important information, as did an interview of Cosme Chacon, Paco's brother, by Leigh Ann Hunt of the Forest Service (Hunt 1994).

It was normally virtually impossible to successfully conduct and record formal oral history-style interviews with Paco. Usually other people were present and drinking. The atmosphere was typically raucous, particularly when his younger brother, Presciliano (Press), was present. Paco was commonly quite reticent, and it was often difficult for me, a non-Spanish speaker with diminished hearing, to understand his English, which was heavily inflected with his native New Mexican Spanish and often noisily interrupted by Press. There would commonly be some English and Spanish or a pidgin form of them being spoken quite rapidly with many interruptions.

Much of this book has come from the author's memories of conversations that were usually held while driving on trips with Paco. These occasions were about the only time we could be alone together. Miles often quietly passed before Paco began pointing out things he saw along the road and talking about all manner of things. It was not possible to record these conversations on tape or in notes. Many of Paco's friends and acquaintances also offered bits of information over the years. Family members were also important sources, but little of this was in written form.

The information contained herein reflects the author's best effort to bring together the materials gathered from all these sources over the past quarter-century. Individual facts about Paco presented here came from all these sources and are not all individually referenced. Along with the rest of my professional research files, the source materials for this project are currently scheduled to be permanently archived in the Special Collections Department of the Penrose Library at the University of Denver in Denver, Colorado.

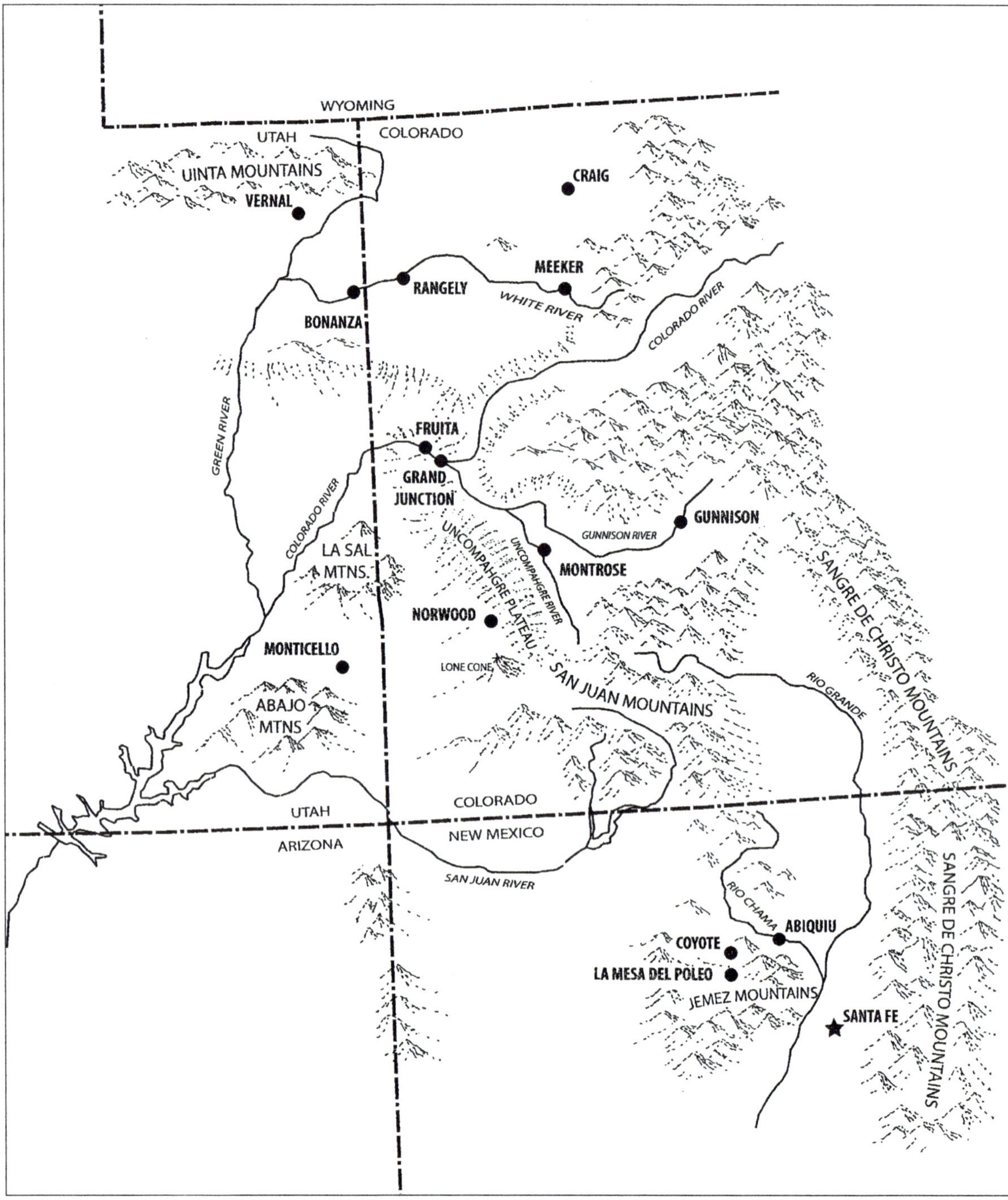

Figure 2: Map of northern New Mexico, western Colorado, eastern Utah, and the places and locales that were important in the life of Pacomio M. Chacon, the sheepherder and master artist of the cliffs and trees.

Chapter 2

HOME — LA MESA DEL PALEO

No one knows when the first New Mexican settlers began living in the northern foothills of New Mexico's Jemez Mountains in the vicinity of the old community of Coyote (founded 1862). It was most likely not much before the middle of the nineteenth-century after hostile Indian threats had diminished among the regional settlements. Only then was it finally safe for people to begin settling in outlying areas farther from the protection of the settlements of the Chama River Valley, such as the Pueblo of Santo Tomás de Abqiuiú some twenty miles to the east.

The early settlers around Coyote included members of the old New Mexican Chacon lineage. Pacomio Chacon's grandfather was purportedly killed in an Indian attack in this region during this period. Some members of this Spanish-speaking family eventually settled back in the foothills south of Coyote. This area is today part of Rio Arriba County in north-central New Mexico. A tiny, dispersed, mixed pastoral and agricultural community developed there, and its members referred to their home as "*La Mesa del Poleo*" or often just "*La Mesa*" (Figure 2).

The name "Poleo" was taken from the Pennyroyal plant (*Mentha arvensis*), which is a form of native wild mint that was plentiful there in the rugged foothills of the mountains (Hendricks 2013). While no town as such ever developed at La Mesa, the church of Santa Teresa was eventually established there and along with a school became the focal point of the little community (Figure 3). The settlement of Coyote was the local supply center for the residents of La Mesa.

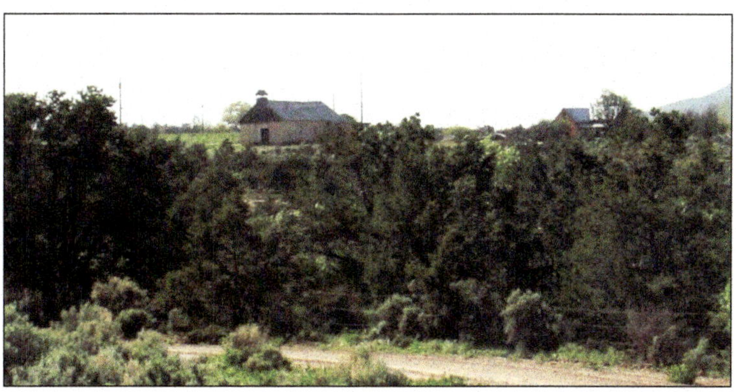

Figure 3: The church of Santa Teresa at La Mesa del Poleo near Coyote, New Mexico, was built in 1946. View is to the northwest in May 2004.

The life story of Pacomio Martinez Chacon commenced in the tightly knit little Spanish-speaking world of the poor subsistence farmers and ranchers of La Mesa. For them a trip by wagon or horseback just as far as Coyote, only twelve miles away, was a big, all-day adventure. Except for going to find work as sheepherders, there was seldom any need for them to venture much farther from home than Coyote. All their simple material and subsistence needs could be met in Coyote or by mail order catalog.

As the third of six sons born to Antonio Maria (1878-1937) (Figure 4) and Elisa Martinez Chacon (b. 1888) at La Mesa del Poleo, Pacomio Chacon was truly a child of that little community. Pacomio is, even today, a name rarely given among Spanish-speaking people. The name derives from Saint *Pachomius* (aka *Pacome* or *Pakhomius*), a Roman soldier of Egyptian birth (A.D. 292-348) who converted to Christianity and became the founder of the spiritual, communal, monastic life (Catholic On Line 2013). Antonio and Elisa were residents of the Coyote area who descended from early New Mexico colonial stock. They married in 1904.

Antonio and Elisa's first child, a daughter, Eufemia, died in early childhood in 1906. The subsequent Chacon children were Jose (November 28, 1908), Cosme (May 10, 1913), Pacomio (April 14, 1916 [aka "Paco"]), Manuel (October 19, 1918), Presciliano (June 3, 1923 [aka "Press"]), and Jesusita (August 26, 1926 [aka "Jesse"]). The Chacon children also had a half-brother, Nastazcio, who died as a young man in an automobile accident. Together with their own children, the Chacons also raised a local orphan girl, Ophelia Martinez.

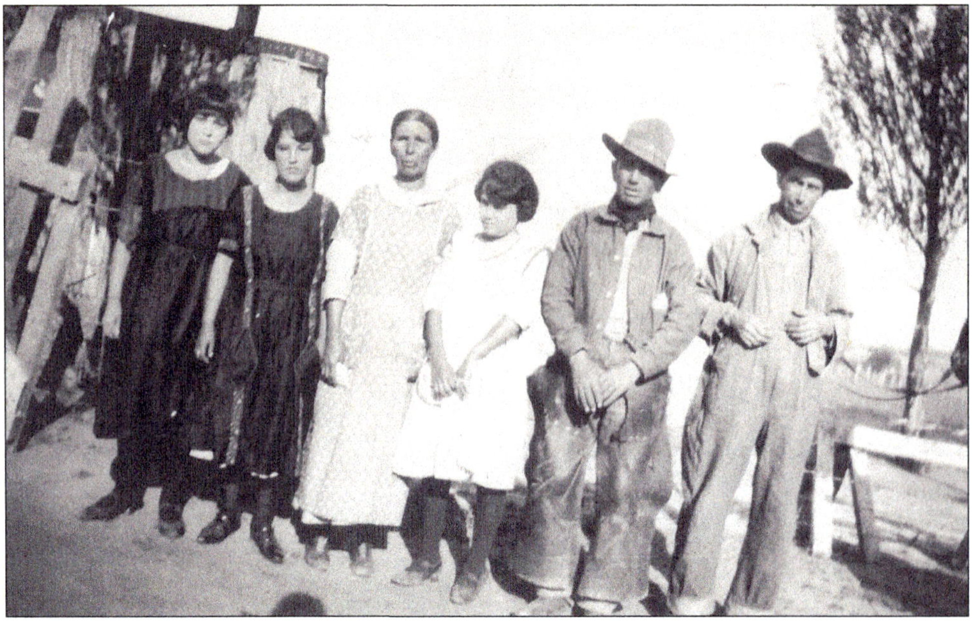

Figure 4: Undated early photo of Pacomio Chacon's father, Antonio M. Chacon, in chaps; Antonio's brother, Nastazcio; Antonio's mother, Maria de Jessus Olivas; and three unidentified younger women presumed to be Antonio's sisters. Photo most likely taken ca. 1900-1910 in the vicinity of La Mesa de Poleo, New Mexico (courtesy Alice [Chacon] Montano, Grand Junction, Colorado).

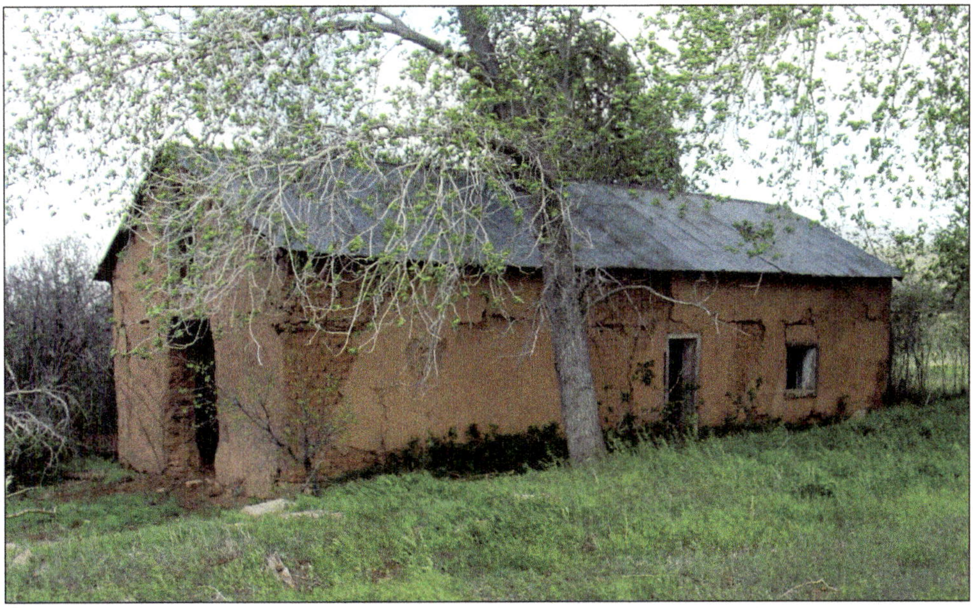

Figure 5: The Territorial-style adobe brick home that Antonio Chacon is believed to have built for his family at La Mesa del Poleo sometime prior to 1916. Pacomio Chacon was raised in this house. Sadly, the end of this lovely old house has been knocked out so that cattle can now use it for shelter. View of May 2004 is to the southeast.

Antonio and Elisa settled their growing family in the rural countryside a few miles south of Coyote at La Mesa. The family farmed its ground and herded sheep and goats. Although there are hints of some brief prosperity for the family, if this was certain, it all changed with the advent of the Great Depression. The Chacon Family was typical of so many rural New Mexican families of the time with lots of children, strong religious faith, and precious little money.

Paco's family lived simply in a substantial Territorial-style house on top of the Mesa del Poleo. This classic structure was of adobe brick construction with three rooms and a big loft, supported by massive log vigas, where the children slept. According to family members, Antonio probably built this house (Figure 5).

Even though he regularly kept a herd of goats to help sustain his family, Antonio often herded sheep for others to earn cash. His sons were all required to help him, and a common wage was thirty-five dollars per month for the services of the father and his boys. Paco was thus born and raised in the world of the sheepherders. This was one of the world's most ancient and honorable professions and long a mainstay in the largely pastoral economy of New Mexico. His family's early rural lifestyle was, by today's standards, almost incomprehensively simple and revolved around his family and the remote little community of La Mesa. There time moved slowly, and things had not changed all that much since its founding.

Paco attended grade school for a few years in a tiny rural schoolhouse near his home. Although he learned the basics of reading and writing, he did not care for school. Later in life Paco admitted that he hated the only teacher he ever had. The term "hate" was not a common part of Paco's vocabulary except when it came to his lack of appreciation for coyotes, the one true bane of all sheepherders. But he used it easily in reference to this early influence in his life. This bilingual teacher was also a man of old New Mexican descent, and Paco believed that he was betraying his heritage by requiring the students to translate their native Spanish into English during their writing exercises. Paco ultimately came to admit that these lessons had been good learning exercises that helped him in later life. During his school years, Paco's artistic talents began to emerge, and he enjoyed doodling, drawing, and copying pictures from books and catalogs.

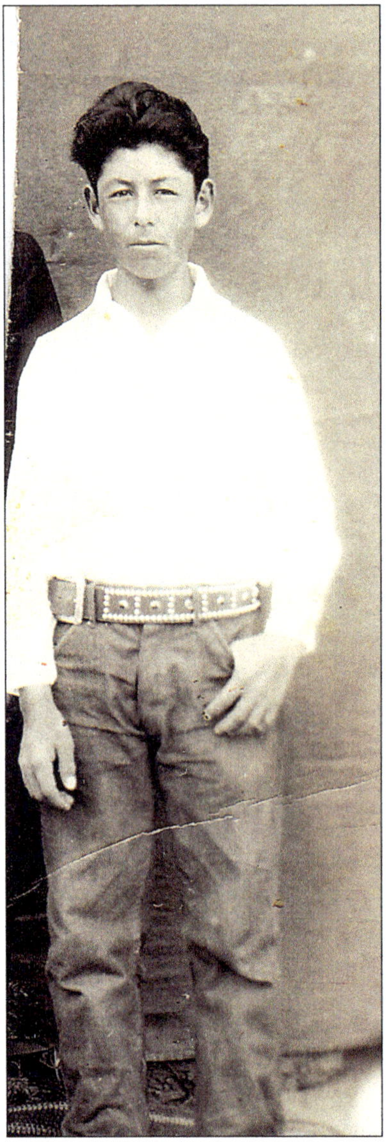

Figure 6: Undated photo of Pacomio Chacon as a young man, seemingly in his teens. This is the only known surviving photo of Paco while he might still have been living at home with his family at La Mesa del Poleo (courtesy of Alice (Chacon) Montano, Grand Junction, Colorado).

Although he really enjoyed drawing, school and work at home kept Paco so busy that he did not have much time to begin refining his artistic talents. The need for him to help his father with farming and herding curtailed the young boy's schooling. He dropped out of school in about 1926 while in fourth or fifth grade. At roughly ten years old, he began working fulltime helping to herd sheep. As he later recalled, his childhood was mostly just work, although he loved drawing and being around his family as well as the sheep and goats that he later came to loath, at least for a time. During his early years herding in the Coyote vicinity, Paco began to carve his name on aspen trees from time to time. He observed that other herders, including family members, often carved their names on the trees and began to follow their examples.

The key to success as a herder is to be vigilant in protecting your charges. His father drove home the importance of this responsibility when Paco was still at home under his roof. Paco, who by then must have been a young teenager (Figure 6), and his little sister, Jesusita, were tending the family goat herd. Paco and Jesse decided to leave the goats in a thicket and go fishing in the creek. When they returned home an old dry doe intended for the family stew pot, was missing. When his father inquired about the missing goat, Paco could not tell him where it had gone.

Antonio retraced his children's steps and found where a coyote had killed the goat while the kids were messing around in the creek. He was furious over the loss of the family source of protein and hauled Paco to the adobe bathhouse that stood outside the family's big adobe home. He beat Paco unmercifully with a razor strop and cursed him, screaming that "he was of no use to the family" if he was going to be so irresponsible. Paco replied that he would just go "shoot myself." His father's response was for him to "go ahead and shoot yourself" since "you are no longer any use to me." He told him "I have enough sons, I have the other ones." Antonio then began to haul little Jesse into the bathhouse to give her a good licking also. Just as he grabbed her, the young child, who must have been only about five or six years old, responded, "Daddy don't hit me, I have to pee!"That was enough for the infuriated father. Jesse's pleas subdued him and he allowed her to go unpunished.

Paco and his younger brother, Presciliano, never forgot this story. By modern measure Antonio's actions appear to have been quite harsh if not criminal by today's legal standards, but there was almost certainly a practical reason for his anger. The impact on the family from the loss of a single goat was probably far greater than we today can even begin to comprehend. Like so many poor settlers in the years of the Great Depression, that dried up old goat might well have been intended to be the primary source of meat for the entire family for a week or more. Life for pioneer settlers like those from La Mesa could certainly be harsh, and an old goat or a flock of chicks were treasured commodities that simply had to be protected.

Paco and his brother frequently reiterated this Chacon Family story to me over the years. Even as an old man, a brief look into Paco's eyes made it easy to tell that he was still stinging from his father's physical and emotional abuse in this one instance. People can easily carry such scars for a lifetime. Paco never forgot the hurt and humiliation he felt that day back in about 1930 at his father's hands. He never abandoned the lesson and believed it helped him to always remember to be vigilant in his responsibility for his charges. It helped him become the fine and conscientious herder that he was ultimately well known to be. Later in life Paco tempered the story for this writer by stating that he had been a "good boy too!"

> *When I was grown I learn to respect my dad and my mom. My father got sick. Then I went and support all the family. Support him. My older brother, my sister, my mom. They don't give us [food] stamps in that time.*

By 1931 the Depression was strangling the American people. All over the country poor settlers of every color and complexion were, as told so well in *Grapes of Wrath* (Stienbeck 1939), finding it harder and harder to keep their families together and fed, let alone be able to hang onto their property. Like the farmers of the Dust Bowl, the poor settlers of northern New Mexico were particularly hard hit. The Chacon Family was no exception, and it became necessary for Antonio to leave his family at La Mesa and go to work as a herder for the Jay Redd Ranch at Monticello, Utah (Figure 7). The father then sent for his son and at barely fifteen years old Paco was also working for the Redd Ranch. Where the pay for a herder had once been thirty or thirty-five dollars per month, the rate during the Depression years fell to only twenty-five dollars and poor and slim rations.

The first New Mexicans from the Coyote area had begun herding sheep in the Monticello area in about 1895. By the time Paco arrived in the area, some of his older brothers, including Cosme and Jose, as well as other men from the Coyote vicinity, were also working in that region as herders. Chacons were then herding on the Elk Mountain Range near Monticello and some had, by at least the early 1920s, apparently been leaving their names on the aspens in that area.

During the summer of 1932, Paco also began herding sheep in this range, and in the winter he worked with the sheep at the main ranch.

Paco was initially supposed to be working as a camp tender responsible for moving the herder's camp to new locations as the herds followed the grass. At barely five feet tall, if that, Paco was so small that he could not reach up to hang the heavy packs on the animals' backs. He had to settle for being only the primary sheepherder's helper who did the busy work but did not reap the wages of an experienced herder or camp tender. Still, it was a useful learning time for Paco, and he eventually emerged with a reputation of being one of the best sheepherders in the entire regional industry. The herders were not generally provided with horses during this period and had to walk everywhere. They did use a lot of burros and mules to carry supplies and move the camps.

While alone in the tent with the primary and much older herder, who is thought to have been his father's brother, Nastazcio (Figure 4), Paco received sage advice. The older man explained to Paco that he simply had to master writing his own well-formed and distinctive signature. As he stressed, it should be a matter of pride for Paco to develop his signature. Otherwise, he would just be thought of as another poor, uneducated herder who could barely write his name. Paco took his uncle's advice, and by the light of a stubby plumber's candle in his bed at night, with the coyotes howling about the camp, he labored to perfect his cursive and printing skills.

These efforts ultimately paid off, and Paco joined the ranks of what he called the "pretty good" old New Mexican "tree writers." This was the term by which the New Mexican herders referred to those who left their names and pictures on the aspens. Late in life Paco considered himself to be one of the last of the tree writers. He commonly demonstrated his considerable abilities both on cliff faces and aspen trees. He was able to leave his distinctive and artistic moniker over a wide region of Colorado, Utah, and just possibly parts of northern New Mexico and southern Wyoming as well. It also served him well while he was in the Army because his comrades would give him cigarettes and candy in exchange for addressing their letters home to girlfriends and family in his beautiful cursive, with the occasional addition of flowers or birds on the envelope.

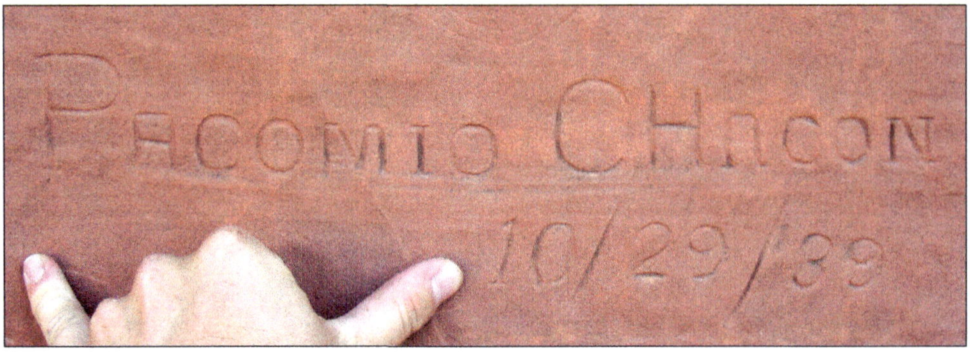

Figure 7: The earliest known example of the surviving, finely executed cliff-face block-printed signatures of the artist, Pacomio M. Chacon. This example is from Looking Glass rock near Moab, Utah, and has recently garnered interest in Pacomio via the internet (photo by Matthew S. Baker).

Although Paco had carved a bit on the aspens in the vicinity of La Mesa, it was not until 1932 while he was working on Elk Ridge near Monticello, Utah, that he routinely began to carve his name. Yet, he had seemingly not begun to carve images of women and animals that would become his trademark. Just when he started to do this is unknown. Paco always laughingly recalled how his youthful time with Redd Ranch was simply walk, walk, walk as he tended six hundred head of sheep or more without a horse. Back then he and the other herders even had to pack salt for the sheep on their backs.

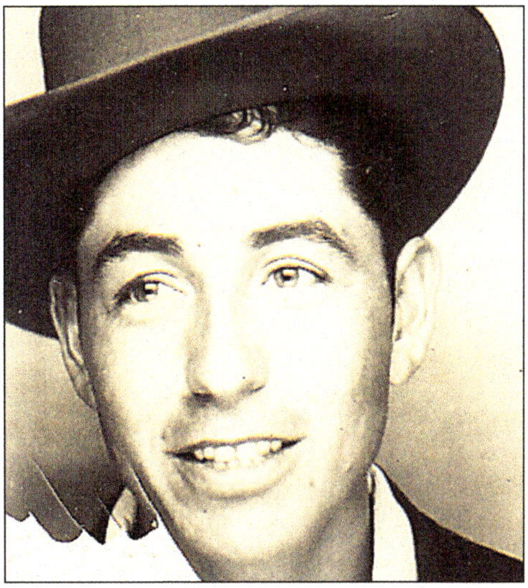

Figure 8: Undated photo of the young Pacomio Chacon ca. 1930s, seemingly from a drugstore photo booth (courtesy of Alice [Chacon] Montano, Grand Junction, Colorado).

Paco and his brother Press equated their lot in those old times to that of the Chinese coolies who labored to build the railroads of America. Paco recalled that the employers essentially made slaves of both the New Mexican herders and the Chinese laborers. He also said the food was poor in those times, but he had no choice since everyone was broke, and a man had to take whatever he could get. Finally, he was earning twenty-five dollars a month working as a herder for Redd Ranch.

Paco worked for the Redd Ranch until 1935 when at age nineteen he went to work for Fred Levi of Millford, Utah. He worked there until 1937 when he decided to seek greener pastures in California. Sheepherding is arduous and lonely work. Many a good man has come to loath it as was so poignantly memorialized in the body of poetry characterized by the well-known piece by Curley Fletcher named *The Sheep Herder's Lament* (Fletcher 1986). The young Paco Chacon was not the first and certainly not the last sheepherder who came to suspect that there must be something more waiting at the end of his personal rainbow than just a flock of sheep. It was at this time in his life that Paco admitted that he was sick of sheep and sheepherding. This idea is reflected in the following lines from the *Sheepherder's Poem* by Mike Lucey.

...They put me with a band of sheep and said you're on your own
and many a time I cursed the day that I left my native home.
Well for eight long years I stayed out there through weather wet and dry
to ward off the hungry coyotes I slept beneath the sky
in my blanket wrapped by a campfire lone
with my saddle under my head
in a land where men had their back to God
where prayers weren't known or said
I never knew what comfort was
the grub was awful tight

so it was spuds and cabbage I had morning, noon, and night.
The butter, eggs, and bacon indeed were awful slim
so when I came in from sheep camp I was both weak and thin
then to all those sheep men ranchers I said a fond adieu
and headed down to 'Frisco for some other work to do
I knocked round from job to job until I settled down
and I never more will roam the plains outside of Lakeview town.

(from "*Poem*"carved on an aspen tree by Mike Lucey, sheepherder [Lucey 2013])

Like the herder in the poem, early in 1937 the young Paco (Figure 8), who was then but twenty-one years old, decided to abandon sheepherding altogether and seek his good fortune elsewhere.

Chapter 3

FISH AND RICE TO FRUITA

Early in 1937 Paco joined the throngs of other Depression-weary migrants seeking an end to their miseries in the sunny climes of the West Coast. With a ten-dollar train ticket to Los Angeles, he traded Utah and the sheep of the Levi Ranch for the agricultural fields of California. Paco received quite a shock when a newly acquired friend took him to a soup kitchen for a free meal. After filling his empty belly, Paco became concerned about how he was going to pay for it. His friend and those who operated the kitchen found it difficult to get Paco to understand that his meal was free, and there was no bill. Paco, the simple New Mexico mountain boy, was absolutely astounded that there was such a thing as charity. This experience likely gave the youth some reason to believe that California might not turn out too bad.

Although Paco fortunately managed to find work there, things did not work out quite as he was expecting. At first, he found himself canning beets for four dollars a day, which was a pretty good wage for 1937. He then went to work cutting asparagus. Most of his fellow workers on this job were from the Philippine Islands, and they happily subsisted on a traditional diet consisting mostly of fish and rice. Even though he liked his Spanish-speaking co-workers and could physically keep up with them at the hard stoop labor, Paco, nevertheless, became discouraged. Although he grew tired of fish and rice, he liked the life and work. But he could not save much money because even then he was sending most of his funds back to La Mesa to help support his family whose patriarch was ailing by that time.

After his short stint with his Filipino co-workers, Paco gave up on fish and rice and headed back east toward Colorado and home. On this trip he joined a group of what he always referred to as "bums or tramps" who taught him how to ride freight and passenger trains for free. By the time he got back to Colorado by way of the rails through Salt Lake City, he had spent only forty-seven cents on his trip! He was proud of this accomplishment when compared to the ten dollars he had originally spent getting out to California. Paco had obviously not found the end of his rainbow and decided that herding sheep, tough as that job could be, was a better personal fit for him than canning beets and cutting asparagus.

Paco had by then already developed some local reputation as a good sheepherder, and he quickly found work with Steve Herndon of Norwood in western Montrose County, Colorado. He stayed at this job from late 1937 until 1941. During this period Paco stepped up his routine

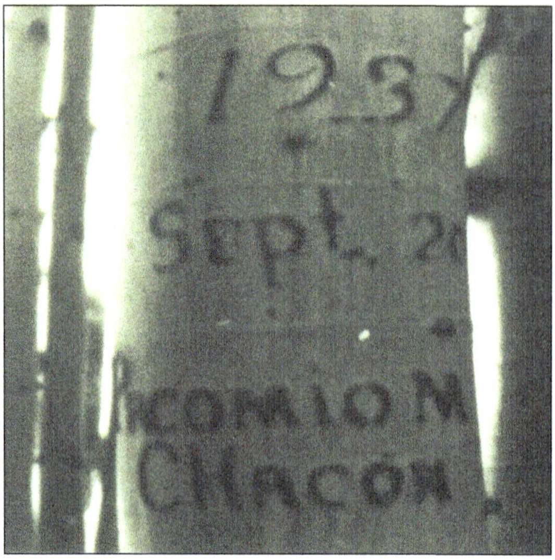

Figure 9: The earliest known example of Pacomio Chacon's carved aspen signatures. This specimen was dated September 20, 1937 and is from the Uncompahgre Plateau in the Lone Cone vicinity near Norwood, Colorado. Paco would have carved this when he was barely twenty-one years old and herding sheep for Steve Herndon of Norwood. It is not believed that he had yet started carving figures on the trees (undated photo courtesy Uncompahgre National Forest, Cultural Resource Files, Delta, Colorado).

of carving his name in the aspen groves (Figure 9) and in the cliffs. To the best of my knowledge, he was still not carving many, if any, women or other figures.

Paco's father, Antonio, died of cancer in about 1937 at age fifty-nine. The newly widowed Elisa found herself nearly alone on the little ranch at La Mesa. Her sons, with the possible exception of the then fourteen-year-old Presciliano, were all grown and off working or serving their country. Elisa still had a young teenage daughter, Jesse, and the young woman named Ophelia Martinez, the orphan from La Mesa, living with her. Although not formally adopted, she had come into the world in 1926, and her parents had abandoned her. Antonio and Elisa took her in and raised her among their natural children as one of their own. After Antonio passed, Elisa moved the three of them to Fruita, Colorado, to be near her own mother. This situation almost certainly played into Paco's decision to leave California. After his father's death, Paco continued herding to help support his mother and sisters.

While working for Steve Herndon out of Norwood, Paco was not all that far from Fruita where his mother, Jesse, and the comely young Ophelia were living. Paco married Ophelia Martinez in Fruita on January 27, 1941, when he was about to turn twenty-five, and she was but sixteen. Paco was, even by this time, known by his family to be overly appreciative of wine and women. Some in the family had also seemed to anticipate that Ophelia would always stay with her foster mother and help care for her in her old age. For whatever reasons, there was, understandably, some considerable "clatter and clap-trap" within the Chacon Family about Paco and Ophelia marrying (Figure 10).

In June 1942 Paco was for some as yet unknown reason in Pueblo, Colorado, where he enlisted in the U.S. Army. He quite likely did so both out of patriotism for his country and to be assured of a steady paycheck to support himself and Ophelia, as well as his mother and sister. He also may have been facing the draft and determined to get ahead of it. Perhaps, like so many other men, he may simply have been coming off a sheepherders' tear or bender that could have altered his senses enough that he went beyond just getting the common tattoo and joined the military. Whatever the reason or reasons, before that year was over, Paco found himself in uniform and receiving monthly pay. He served until November 8, 1945 when he was honorably discharged with a rank of private first class and was earning perhaps twenty or twenty-five dollars a month.

Paco was nervous about going into the Army, and his mother advised him on how to stay out of trouble. She stressed good behavior and avoiding being led astray by others. She imparted old New Mexican folklore by telling him that if he behaved like the coyote, whom all from sheepherding families hated, he would "begin to howl like one." She told him coyotes steal but that good boys should never steal like the coyote. She also said "please remember me, I don't steal" and instructed him to unquestionably obey any orders he was given.

Although he eventually came to an accommodation with military life, Paco's introduction to it was not easy. In his own words he went into the service "not knowing nothing."

> *I never been with soldiers. We were raised in the mountains. I went and to me was strange. Everybody call me all kinds of names. I don't care because I don't talk good English...they come from Virginia, they come from Missouri. I never even know Missouri got people there, but they only think I write good. I make letters for them, and then I went in, he [officer] wanted me to put all the names for the count [roll], cuz there were troops coming. And all went against me. How they make you a sergeant or corporal when you don't have no school? I*

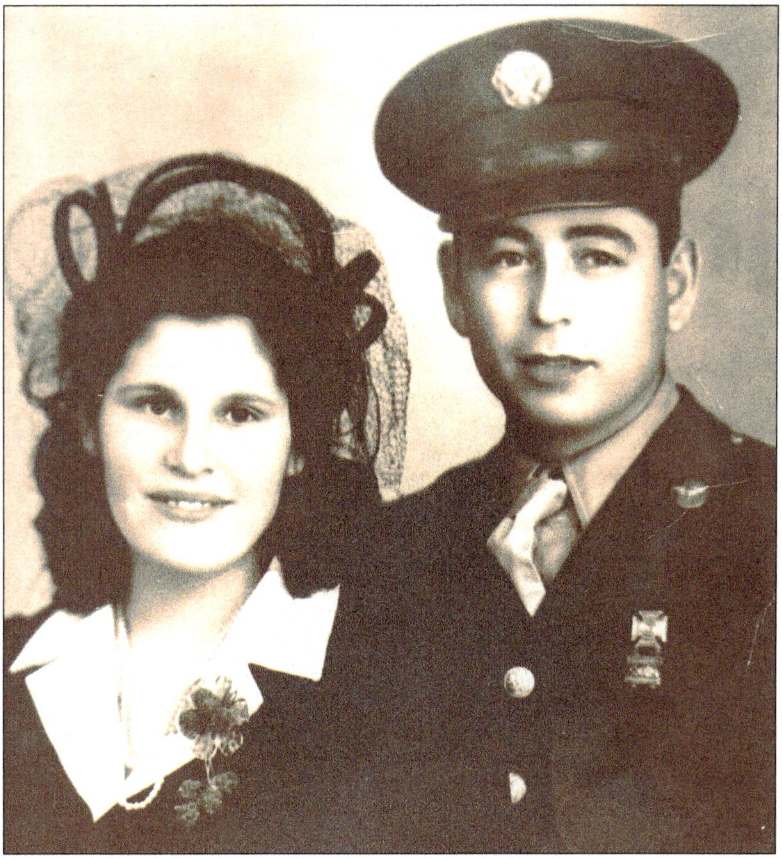

Figure 10: Undated photo (ca. 1942) of the newly wedded Ophelia Martinez and Pacomio M. Chacon (courtesy of Alice [Chacon] Montano, Grand Junction, Colorado).

told them one thing, my hand is good. I put all the names there. That's about my job. And the officer say he likes me. You don't have to work hard, you can go out and exercise two hours. I hear the other guys, you know what, they say your nothing! I tell him I'm not, it's not my fault, they put me to do that. But a lot of guys want me to put addresses for their girlfriends...

Paco did not see combat and ended up being trained as a cook's helper. This was, however, only after he received some training as a ball turret gunner for the Army Air Force. His diminutive size of barely five feet and perhaps 120 pounds made him a perfect physical candidate for such training. Although he already had some experience with guns, as Paco once explained to me, he was unable to keep up with all of the bookwork required of an aspiring ball turret gunner. He once said that he was told by the Army training personnel that he "had no IQ" and then added "whatever that means!" For all who knew Paco, this lack of formal educational knowledge is easily understood.

Although he was obviously an intelligent man who could read and write, Paco was not schooled in the traditional sense of the word. While fundamentally literate, his schooling was not in math, science, and such but as a sheepherder who had to know sheep and the "Book of Mother Nature" forward and backward. In this regard he always maintained high marks. Yet, his formal and informal education almost certainly did not expose him to the finer points of understanding the meaning of intelligence quotients or the geometry of machine gun cross-firing at fast-moving enemy aircraft!

After leaving gunnery school, Paco remained stationed in Hawaii (Figure 11) for a time and, overall, enjoyed the Army. Although as a cook's helper he likely peeled a lot of spuds and scoured a

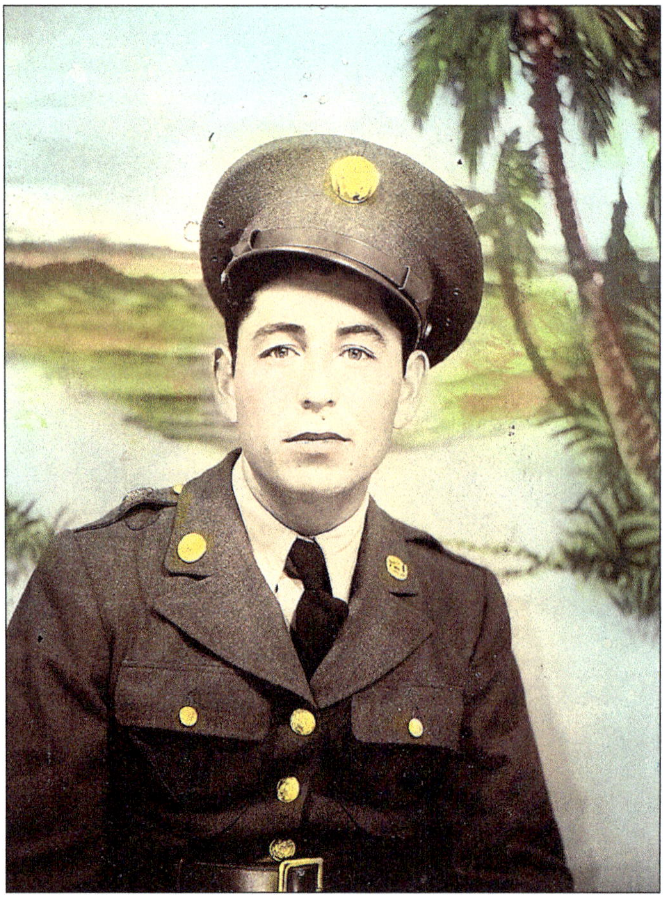

Figure 11: Ca. 1942 studio portrait of Pacomio Chacon taken while he was stationed in Hawaii (courtesy of Alice [Chacon] Montano, Grand Junction, Colorado).

lot of pots, he always had enough to eat, a warm place to sleep, and a steady paycheck. His kitchen duties were certainly much, much easier than those of a sheepherder, and he was not facing the perils of combat. Still, his military pay was probably less than that of a good herder. During the years of the Depression, many good men joined the military; it was then one of the most readily available jobs around. After the attack on Pearl Harbor there was also a ground swell of patriotic sentiment that further induced men to enlist.

The importance of a regular paycheck for the newly wedded Paco and Ophelia would certainly have been a critical consideration. Paco emerged from the Army as a good cook who later on always kept a militarily styled clean and neat camp that was a consistent source of pride for him. The "spit and polish" of the military never left him. It was not unusual for him, when speaking of some other herder, to sarcastically whisper to me that "he no very clean, he no keep clean camp!"

Ophelia remained behind in Fruita during the time that Paco was stationed in Hawaii. Paco was then ordered to a duty station in Wisconsin where he served until his discharge. Ophelia was able to live with Paco for a time during this period of his service. Tragedy struck when they lost at least an infant boy. Ophelia became pregnant again while living in Wisconsin and returned to her mother-in-law's home in Fruita where their first daughter, Alice, was born on April 26, 1944. After Paco's discharge the young couple moved back to Fruita. Paco was then twenty-nine years old and no longer had his military pay to support his little family of three.

With a wife and child to support, as well as his mother and sister, Paco returned to Colorado as just another face in the throngs of newly discharged and suddenly unemployed young American men who managed to come back from the war alive and not maimed. He was among the lucky ones who found work when he got home. He managed to get on with the Denver & Rio Grande Western Railroad out of Grand Junction. He worked as a common laborer, but at least he had a job and, as a child of the Depression, thought he had it made.

Unfortunately, a host of former railroad employees also returned from the war and reclaimed their former positions. After only eighteen days on the job Paco, was terminated because of this influx of men with more seniority. As Paco's brother, Presciliano, once related: "We no raised for railroad. We raised for sheepherding." Once again Paco had to take up the only real profession he knew and went to work herding sheep for Kermit Redd of Blanding, Utah. This work took him into the high summer pastures of the Uncompahgre Plateau in west-central Colorado.

Paco never again attempted to turn his back on sheepherding and worked for many prominent sheep operations in Utah and Colorado over the next fifty years or so as outlined in Table 1. Following a short stint with Kermit Redd, Paco worked around with various sheep operations until 1948. He then went to work for Eva Fitzpatrick of Cottonwood, Utah. He found that Fitzpatrick was a good employer, and he enjoyed his time working for her. This work took him up into the high summer mountain pastures of the San Juan Triangle about Silverton, Ouray, and Lake City, Colorado. Fitzpatrick paid Paco fifty dollars a month but then raised his wage when she had to begin deducting social security. Paco eventually worked his way up to two-hundred fifty dollars a month with Fitzpatrick. He once tried to describe how he started paying social security.

> *That when they to start to get social security out of me. If I [Fitzpatrick] want to help you guys [the herders] this social security she explained to me it gonna cost you six dollars a month and I pay half, I pay twice, government pay three time, and you pay yours one time, and that help you.*

Paco stayed with the Fitzpatrick operation until 1951 when she sold out to Emmet Elizondo of Montrose, Colorado. He stayed on with Elizondo until 1954. Elizondo was a Basque, and this was the first time that Paco had been around the Basque sheep men and herders. He and the other New Mexican herders always referred to them by the term of "Vascos." Paco was never disparaging of the Vascos as an ethnic group, although he might comment on the nature or peculiarities of specific individuals of this group, just as he would of any other ethnic or racial group. Paco and Press thought the Basque herders were quite good at their jobs, and they learned things from them. The Basque herders taught Paco how to make sheepherders' bread in a cast iron Dutch oven set in hot coals. He learned well, and his family still brags about his bread.

Table 1:
Summary of Pacomio M. Chacon's Employment as a Sheepherder

1932-1935	Jay Redd	Monticello, Utah
1935-1937	Fred Levi	Millford, Utah
1937-1941	Steve Herndon	Norwood, Colorado
1941-1945	U. S. Army	
1945-?	Kermit Redd	Blanding, Utah Uncompahgre Plateau, Colorado
?- 1948	Various employers	Western Colorado
1948-1951	Eva Fitzpatrick	Cottonwood, Utah American Flats—Above Ouray and Lake City Area and Mineral Point at Silverton, Colorado
1951-1954	Emmet Elizondo (bought out Fitzpatrick)	Montrose, Colorado Big and Little Cimarron River, Blue Creek, Sapinero, Colorado
1954-1959	Dave and Perry Christianson	Grand Junction, Marble, Colorado
1959-1972	Louis Livingston	Craig, Colorado Craig, Colorado & various ranges
1972-1978	Mike Theos	Meeker, Colorado Rangely & Meeker, Colorado, and Wyoming?
1978-1981	John Perulis (bought out Theos)	Meeker, Colorado?
1982	inactive as herder	
1983-1985	Seeley Ranch	Meeker, Colorado seasonal only
1985	inactive as herder	
1986	Squire Ranch	Meeker, seasonal only
1986	occasionally helping friends and relatives with lambing and short-term work	
1994	fully retired except for very occasional work	

Paco and Ophelia settled in Fruita, and Paco began a lifetime of sheepherding and being a largely absentee husband and father to a growing family. The Chacon's ultimately had six children whose birth order is: Alice, Suzie, Sophie, Pacomio Jr., Marcella, and Priscilla. At this time Paco seems to have initiated his artwork in earnest while tending his various flocks. The stresses of the sheepherding life played hard on the Chacon Family. All the normal challenges were there and complicated by Paco's drinking and womanizing. They finally led Ophelia to completely wash Paco from her life. The couple separated in about 1957 but never divorced. During the emotional turmoil of this separation, the sad and demoralized Ophelia burned or threw away nearly all the family photos and archives, including much of Paco's artwork.

Ophelia died from cancer in 1982 and was buried in the Fruita Cemetery. Paco never remarried. When he was not out with the sheep he might stay awhile with a girlfriend of the moment or his daughter, Alice, at her home in Fruita. Alice cared for him when he developed Rocky Mountain Spotted Fever and was unable to work for about a year. It was largely through Alice's efforts that he realized some periods of sobriety when not out herding sheep. Still, it was not uncommon for him to turn into another of the "crazy sheepherders" and go on a "tear." While he was actively herding he generally did not drink. Even if he had wanted to, it was difficult to get much liquor into his camp. He retired from sheepherding in 1994 but still helped friends and family members on special occasions such as during lambing season.

Through the late 1990s Paco lived with his younger brother, Press, in a small trailer home on East Carolina Avenue in Fruita. In about 2000 he moved to The Oaks, an assisted living facility in Fruita. In his last years, he developed Parkinson's Disease and then fell and broke his right hand. His shakes from the Parkinson's and this injury kept him from effectively ever again doing any form of decent artwork as told herein in Chapter 10.

In his last years, Paco developed dementia and lived in a locked ward known as "Main Street" at Family Health West in Fruita. His family, the staff of the nursing home, and I posted pictures of sheep and examples of his artwork along the halls as a means of trying to lead him back to his own room. Paco died there on July 21, 2009. His services were held at the First Assembly of God Church in Fruita on July 25, 2009. His ashes were later interred in the Veteran's Memorial Cemetery in Grand Junction, Colorado, with full military honors.

Chapter 4

TO BE A SHEEPHERDER

In past generations there were two basic kinds of sheepherders. Some, like Paco Chacon, were just regular people born into the profession, and it was the only lifeway they ever knew. In this regard they were like many other people, including farmers, carpenters, and a host of others who followed in their family's occupational footsteps. Others came into it as adults from a host of different directions and for a variety of reasons. No matter what their backgrounds, virtually all herders eventually come to either love or loath their profession. Some, like Paco, ultimately learn to love the sheep and are accepting of the sheepherding lifestyle. This may be because they never had an opportunity to do anything else.

Others never stayed at it and really learned to love their charges. The latter have given rise to all kinds of negative connotations about the lifestyle, which are readily found in such works as the well-known *Sheepherders Lament* by the famed cowboy poet, Curley Fletcher. Although he boasted of his exploits as a cowboy and wanderer of the American West, Fletcher emphatically stressed that he had never fallen so low as to have had to herd sheep (Fletcher 1986:8).

The Spanish Inquisition might
Have been a whole lot worse,
If instead of crucifixion, they
Had some sheep to nurse.

Old Job had lots of patience, but
He got off pretty cheap-
He never knew what misery was,
For he never herded sheep.

...It's nice to have your mutton chops,
And your woolen clothes to wear,
But you never stop to give a thought
To the man that put them there.

The blind and deaf are blessed,
The cripples, too, that creep;
They'll never know what misery is,
For they never will herd sheep.
(Fletcher 1986:35)

Sheepherding hallmarked the history of Paco's New Mexico homeland since the founding of the little Spanish colony in the late sixteenth century. New Mexico's multitude of small rural settlements, such as La Mesa del Poleo, were in very large measure developed on the products derived from sheep and goats, as was much of the rest of the American West. A substantial portion of New Mexico's population knew no other lifestyle beyond herding and likely seldom considered their lot quite as miserable and unfortunate as that described by Fletcher.

Perhaps next to old-time lighthouse keepers and fence-riding cowboys, sheepherding is certainly among the loneliest of professions. In keeping with Fletcher's characterization, it has long been popularly cited as the deciding factor in bringing about all sorts of mental peculiarities in those who have followed its ways for very long. The term "crazy as a sheepherder" has joined "mad as a hatter" as part of the popular American lexicon. Writing about the sheepherding life back in the 1920s (Figure 12), one old-time herder spoke of this characterization with tongue-in-cheek and outlined his two general theories about sheepherding.

Some hold that no man can herd for six months straight without going crazy, while others maintain that a man must have been mentally unbalanced for at least six months before he is in fit condition to entertain the thought of herding.
(Gilfillan 1993:3)

Many who were born to the sheepherder's lifestyle remained mentally healthy, fairly regular people throughout their lives, even though they never departed from the ways of herding and/or sheep ranching. Testimony to this may be found in the way many western sheep ranching family dynasties evolved from immigrant herders who came to the United States. There have certainly been individuals who were so solitary in their nature that they chose sheepherding over other better employment options simply for the opportunity to escape the presence of other people.

Others have been known to take up herding to wean themselves from alcohol or even to disappear from the radar screens of law enforcement, ex-wives, and obligations for child support and life's other complications. Like Paco Chacon, many others had no overt problem with people but knew little else. They never had the opportunity to cast the profession away and adapt their personal skills to any other gainful pursuits. That was certainly the case for Paco.

He not only had very limited education and no other marketable skills, he also had come to manhood during the Great Depression when options for making a living were simply not available, even for the educated and highly skilled. As one old herder named Cad once stated about the Depression, 1932 was a time when there simply "wasn't a nickel in the world (Mathers

1975:24, 12)." For men like Paco, who started herding during the Depression, a dollar a day at best did not seem like bad money, even if it was then popularly perceived to be one of the lowest jobs anyone could hold. Had it not been for his artwork, Paco would, at his death, likely have just disappeared from public memory along with the multitudes of other now forgotten herders who passed before him through the millennia.

Before discussing Paco and his art any further, I would like to acquaint the reader with the profession of sheepherding as a lifeway and explain its many demands and simple rewards from the perspective of the herders. The discussion is not framed from the perspective of the sheep ranchers who employ the herders. As explained to me by one such individual (Etchart 2014), there are two sides to every issue, and the herders and their employers have not always seen things in the same light. By explaining the industry from the herder's perspective, the reader should begin to understand how Paco came to be an artist of the cliffs and trees and why he carved what and where he did.

The reader will also note that although Paco at one point in his youth came to loath his charges, by the time he was an old man, who was admittedly not deeply religious, his respect for and love of sheep had become nearly biblical in its sensitivity. In contrast to those who chafed at their profession, a few of the old herders disliked being called sheepherders and thought of themselves more as the shepherds in the biblical descriptions. Writing in 1992 one retired Utah herder spoke as follows:

> *I consider myself a shepherd. There's nothing really wrong with being called a "sheepherder." It's just that "shepherd" seems like a stronger word. It better describes what I do and the way I feel about myself and my sheep. A shepherd is a guide, a leader, a guard... a person with a flock that he really cares about.*
>
> (Crook 1992:1)

As far as I understand it, there has historically been a practical distinction between sheepherders and shepherds. This distinction involves the method used to move the sheep between destinations. The sheepherders have tended to drive the sheep from behind. They commonly use dogs to nip and bark at their heels to keep them moving before them. By contrast, shepherds tend to lead their flocks from the front. They invite the trusting sheep to follow where they lead. The latter description seems best suited to biblical descriptions of shepherds.

> *My sheep listen to my voice: I know them, and they follow me.*
>
> (Gospel of St. John 10:27)

Sheepherding is a lifeway and man's oldest organized industry (Paul 1976). It is not a profession you can take up lightly. The profession began in the depths of human history, perhaps as far back as 10,000 years ago, when sheep were first domesticated in Asia. It may well be the earth's oldest profession and factors into the very core of biblical iconography and literature in names such as those of Abel, Abraham, Moses, and King David. It is central to the lovely and ever reassuring Twenty-third Chapter of the Book of Psalms. Even the prophet Mohammed had been a shepherd.

Shepherds lie at the very core of Christianity, as they were the very first people to see Jesus, the veritable Lamb of God, after his birth. It is thus not surprising that sheep are mentioned more times in the Bible than any other animal, at least five hundred times. The followers of Christ, their shepherd, have long been described metaphorically as his flock of sheep (Schoenian 2013) as embodied in the previously outlined distinction between sheepherders and shepherds. In the United States the term "sheepherder" long ago generally supplanted that of "shepherd."

Although the name has changed, and the profession has taken on a more commercialized and secular nature, those who tend flocks of sheep still, despite technical innovations and whether or not they lead or follow their flocks, live a lifestyle that has changed little since biblical times. As the ultimate in domesticated animals, sheep are completely defenseless and dependent on man for their protection and succor. They will not even fight to defend their young. The fundamental duties and obligations to the flocks have remained the same for thousands of years. This is clearly revealed in another look into the Twenty-third Psalm wherein the dependence of believers is equated to the sheep's dependence on their herder.

> *The Lord is my Shepherd, I shall not want.*
> *He maketh me to lie down in green pastures:*
> *He leadest me beside the still waters…*
>
> (Gospel of John, 23rd Psalm)

Although there were once many American-born Hispanic, Greek, Basque, and other sheepherders like Paco Chacon working throughout the western U.S., they have today in large measure been replaced with foreign workers brought in from Peru, Boliva, and other South American countries on temporary work permits. Some are now even brought in from as far away as Tibet. Even though the backgrounds of the herders have changed dramatically over the last half-century or so, the nature of the work has not. This is despite the disappearance of the sheepherder's tent, old style sheep wagons, pack mules, and the advent of solar powered sheep wagons, four-wheel drive vehicles, and modern communications.

A herder's duties almost always take place outdoors in all kinds of weather, no matter how brutal. This fact of a herder's live has never changed and will remain an axiom of the industry so long as it survives. A 2013 job listing for sheepherders in Colorado does not differ much from how it would have read many decades ago.

> *Position: Sheepherder Duties: Tends flocks of sheep grazing on range or pasture; moves sheep to and about area assigned for grazing; prevents animals from wandering or becoming lost, using trained dogs to round up strays and assist in moving flock to other locations. Beds down sheep near campsite each night. Guards flock against predatory animals and eating poisonous plants. May assist in lambing, docking, castrating, dehorning, shearing, vaccinating, drenching,*

and medicating animals. May attend sheep and lambs in barns during lambing season. May brand, tag, clip or otherwise mark sheep for identification purposes. May sort and cut culls. May feed animals supplementary rations. Workers must be willing to perform all duties according to the employer's specifications during the entire contract period. Workers will be on call 24 hours a day seven days a week as needed. Workers will be expected to work in conditions normally associated with Colorado climatic conditions, which may include at times dusty conditions, wind etc. Temperatures in early spring and late fall are cool; Winter cold and temperatures during working hours can reach 100 degrees Fahrenheit. Skills Required: Must be able to ride a horse. Must be able to repair fences. Must have at least three months experience as a sheepherder. Other Requirements: must be 18, for labor law requirements; Must have at least one employer reference verifying sheepherding experience. If employment was more than 5 years ago 2 employer references must be provided. Wage: $750/month. Shifts: days shifts but will work long hours as needed (on call 24/7) 3/15/2013 till 3/14/2014 contract period). Job Site: Jackson County area in ranch setting and remote locations. Benefits: Room and board....
(Wyoming Jobs 2013)

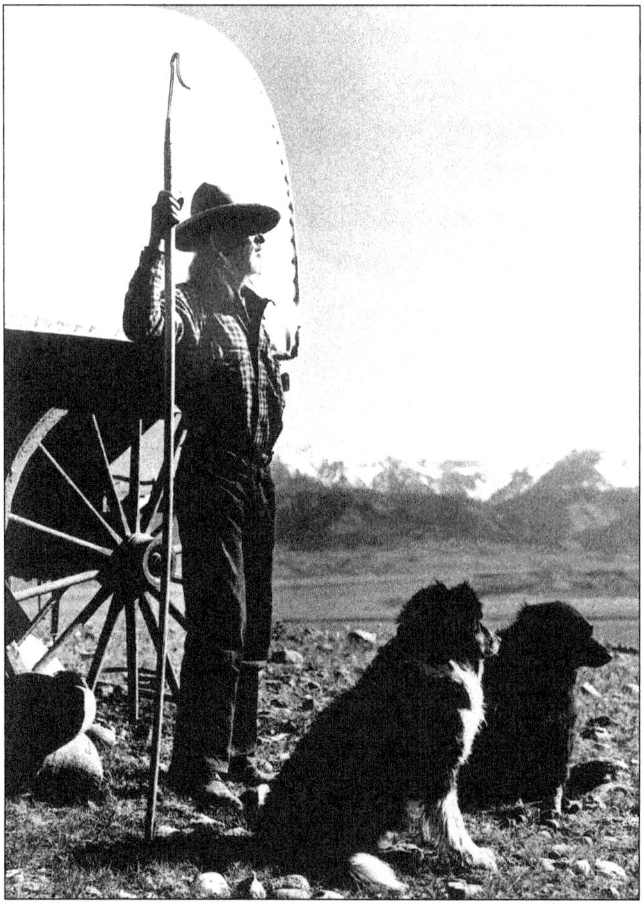

Figure 12: Undated late nineteenth- or early twentieth-century photo of a sheepherder, his dogs, and sheep wagon in a mountainous area somewhere in the western U.S. The herder holds a shepherd's hook, which is used to catch sheep by the leg (courtesy Western History Department, Denver Public Library No. F19059).

Sheep ranchers (Paul 1975), the men who actually own the sheep and employ the herders, normally provide them with their board and a portable "sheep wagon," which is sometimes called a "sheep camp," in which to live. Sheep wagons are a now traditional form of trailer home designed to

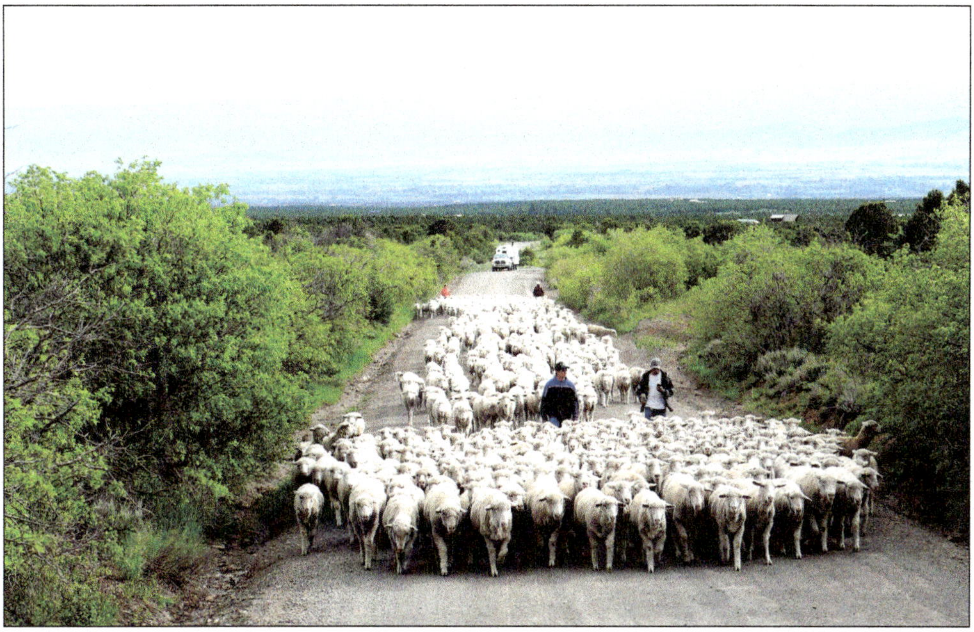

Figure 13: Vast herds of sheep being trailed to summer pastures once commonly blocked roads of the American West. This northerly view of June 2014 shows a flock of sheep being trailed in this now rare, traditional manner by the Ernie Etchart family of Etchart Ranches in Montrose, Colorado. This flock, followed by the sheep wagon home of the herder, is being pushed up the Dave Wood Road to spring pastures on the Uncompahgre Plateau in Montrose County. The Uncompahgre Valley and Grand Mesa are in the background (photo opportunity courtesy of Etchart Ranches, Montrose, Colorado).

meet the specific needs of sheepherders on the vast ranges of the American West (Weidel 2001). Even with their groceries and a portable home provided for them, few contemporary Americans would even consider responding to a job offer like that described above for only $750 per month. This is particularly the case so long as there are food stamps, unemployment compensation, and other forms of public assistance. These entitlements all together quite likely can pay more than sheepherding and require little or no work.

It is little wonder that most herders today come from impoverished Third World countries where they do not have such safety nets. To them such a paltry wage seems attractive enough to draw them away from their families for years at a time. The advertised wage has remained largely static for many years, and the work is still every bit as hard as it was in the past. In 1992, nearly a quarter of a century ago, an ad for herders in western Colorado listed the wage as only $650 (*Grand Junction Daily Sentinel* 1992).

In Paco's case there was little to no public charity available at the end of World War II. When you had a family to feed, you took whatever work you could get; there were no other choices. After his stint in the Army, Paco became a herder once again, probably with little realization that it would remain his life's calling.

The old saying about there being a season for all things is certainly an axiom in the sheepherding lifeway. In the Rocky Mountains sheep are gathered in springtime from their winter ranges. Lambs are born, and the flocks are prepared for moving to distant summer

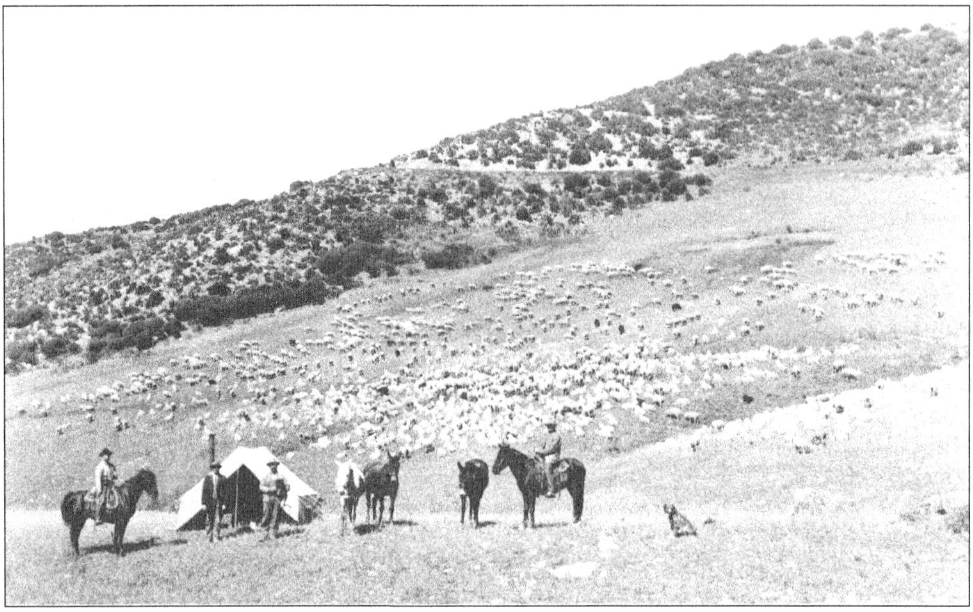

Figure 14: A late nineteenth- or early twentieth-century photo of sheepherders with their flock and traditional tent camp with sheepherder's wall tent, apparently in the vicinity of Cimarron near Montrose in western Colorado (courtesy Western History Department, Denver Public Library No. F46929).

pastures. This preparation involves castrating, tail docking, dehorning, vaccinating, marking, drenching (giving them liquid medicines), and shearing them of the winter's growth of wool to sell at market. In the past the sheep were then driven on the hoof to their summer grazing ranges, which were commonly high in the alpine meadows many days travel away. Sometimes they were first driven to spring pastures at intermediate altitudes to wait until the alpine pastures were snow free (Figure 14).

In the fall the sheep were gathered and driven toward their wintering grounds at lower and warmer elevations. Young, fat, market lambs less than a year old were culled from the herds at that time and sent to market. Today sheep are commonly trucked between summer and winter ranges. The once common sight of vast flocks of sheep blocking major highways followed by a herder's sheep wagon (Figure 13) have all but disappeared from the Colorado landscape.

Sheep require a lot of pasturage. Flock owners obtain rights to graze them on the public lands of the American West through permit systems managed by agencies such as the Bureau of Land Management and the U.S. Forest Service. In past generations the owners relied on herders from families of Hispanics, Greeks, or Basques who came from sheepherding backgrounds and were known to them. Many of the herders in western Colorado and eastern Utah came into herding from New Mexico. A sheep rancher might hire someone such as Paco from the vicinity of Coyote. By word of mouth, announcements of job openings were communicated back to that area, and soon there might be many herders from some specific community or locale herding about a region, particularly if they were good herders like those from the Coyote region.

In the old days, herders were even recruited from selected regions of Europe by the same methods. Eventually these old channels of employment began to break down as more and more of the old American herders found better avenues for achieving the "American Dream" and gave up on herding. This brought about the present system wherein herders are often brought in from places such as South America where herding cannot be as profitable as it is here, even with only the present paltry wages paid in the U.S. These herders are brought in by the sheep ranchers under the Federal H-2A program, which allows for the hiring of foreigners for jobs that U.S. workers will no longer take (Finley, Bruce and R.J. Sangosti 2005).

Even with the availability of the foreign herders, regional sheep ranchers have become a vanishing breed, and the industry is today but a shadow of what it was only a few decades ago. The reasons for this include the advent of synthetic fibers that have replaced wool in textiles, rising land prices, falling profit margins, and the long hours and arduous nature of the industry (Siniai 1999). A herder from Mexico foresaw this situation and left terse messages about it among the aspens of central Nevada in the 1920s. He stated that "No more sheepherders after 100 years" and "Goodby mountain hospital of Sheep (McGonagle 1990:41)."

Chapter 5

PACOMIO THE SHEEPHERDER

Paco Chacon had a reputation as a particularly honest and dependable herder. When other herders were needed in the regional sheep industry, ranchers often came to him and asked if he knew anyone who wanted a job. Paco then wrote or phoned home and asked if an uncle, brother, cousin, or friend wanted a job. This is exactly how Paco got into the local industry. His father, uncle, and other relatives went to work first in Utah, and some remained there for life; Paco followed them. When he learned of a job opening Paco described the country for those who might be seeking work and, most important, he evaluated the employer for them. At one time there were a lot of Chacons, Archuletas, Gallegos, Lovatos, Lujans, and Martinezes from northern New Mexico working in western Colorado and eastern Utah.

Among the herders the evaluation of a potential employer was all important. Some employers were known to be fair, thoughtful, and even generous bosses. Others were known to be just the opposite, and they had significant difficulties obtaining and retaining the better herders. This issue of evaluation of bosses remains a critical component in the industry today and is the subject of frequent labor complaints to the authorities. This is one of the areas where the views of the sheep ranchers and those of the herders are often at odds.

The government has established fair labor standards and a host of rules and regulations for the sheep industry that were not in place in earlier decades. Although they tend to curtail some abuses, they also place significant, at times costly, constraints on sheep ranchers. Foreign nationals, who often do not speak English and are very poorly educated, are so dependent on their employers that it can make it difficult for them to assert their rights under the U.S. and state laws. A bad and stingy employer who does not pay on time or provide good and adequate food, when added on to the routine workload, can quickly make the life of a herder unimaginably difficult (Neal and Humphrey 2001).

Winter and summer ranges were quite different and had to be suited to the seasonal needs of the sheep. Prime summer ranges were often high in the alpine zone of the mountains, and sheep were taken to them as soon as the weather and snow conditions permitted, commonly in July. The sheep were left in the high country in the care of their herder until fall. As the fall weather progressed, the herder would begin to push his charges down from the mountains to a gathering point where the fat spring lambs were culled out for market. The rest of the flock was then trailed or trucked to wintering grounds.

Good wintering grounds were often located in the canyon country where conditions were mildest, and there was feed and permanent water. The warm sunny slopes of deep canyon walls provided good resting grounds for the sheep. When the sheep were still being driven between the summer and winter grounds, the owners had to arrange for stopping points along the way where the sheep could be grazed, rested, and watered. When such stopping places could be found the sheep could be slowly trailed between the winter and summer ranges. To efficiently trail sheep between the two there were limitations on how far apart they could be. Once they began trucking sheep between the summer and winter ranges it was possible to utilize grazing areas that were much farther apart.

For a number of years in the 1970s Paco wintered the sheep of Mike Theos of Meeker on his winter range allotment in the canyons of the Douglas Creek Arch in the locale known as Shavetail Basin. This is about ten miles south of Rangely in Rio Blanco County of northwestern Colorado. Theos once described this rugged region to me as some of the best winter sheep range in the United States. This is where I first came into sustained contact with Paco's art. He left some of his best work in this canyon-ridden area. Although good sheep winter ground, it can be an unpleasant place for people in summer and winter. One individual, surely a cowboy or sheepherder, once left his testament to the region on a rocky cliff face and stated "the reason we are here is because we ain't in hell but we are on our way!" He left this where Paco later

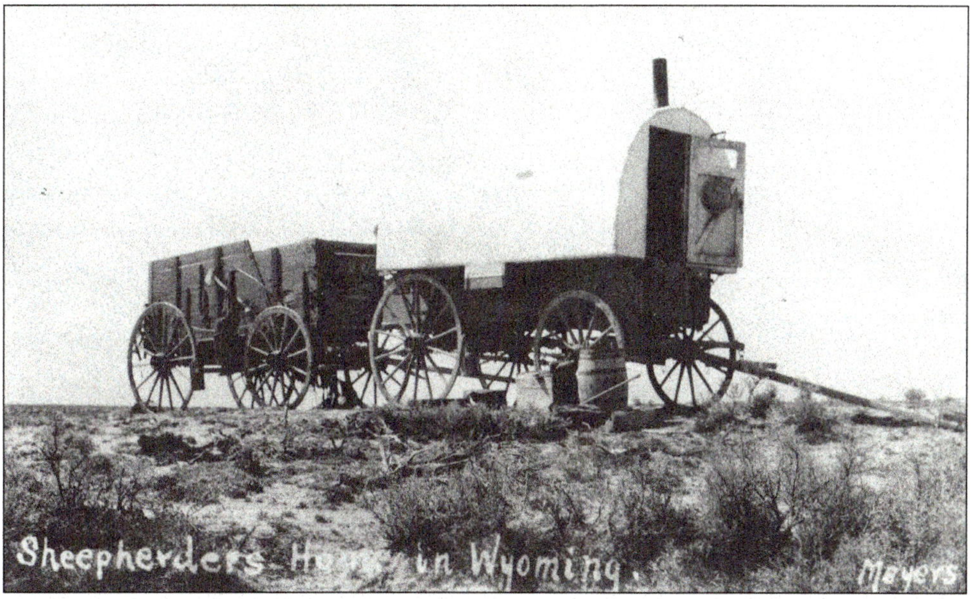

Figure 15: An undated but obviously early, probably nineteenth- century, photo of the old traditional style sheepherder's "hookup" near Rawlins, Wyoming. The photo caption states that this kind of horse-drawn, two-wagon camp unit was invented by James Candlish in Rawlins, Wyoming, in 1884. Weidel (2001:38) concurs with this notion. These old sheep wagons are commonly believed to have been patterned after the classic bow-topped Gypsy wagons of Europe. The front wagon is the "sheep wagon," which was the herder's home. The trailing wagon is the commissary unit also known as a "hooligan." It is used to haul supplies for the herder as well as for his horse, dogs, and the sheep. The definitive history of sheep wagons is given in Weidel (2001) (courtesy Western History Department, Denver Public Library No. F46928).

carved his lovely lady known as his "Dream in Gossamer" (Figure 46; Chapter 8, PRA 9). In summer, Paco tended the Theos flocks on the allotments in the White River National Forest in the high mountains around Meeker, Colorado, roughly sixty miles to the east up the valley of the White River. Paco trailed Theos's sheep between the two areas, and the trip normally took about two weeks. During this period Paco was typically tending a flock of from six hundred to eight hundred sheep.

Good winter sheep grounds were also present in the region about Vernal and Bonanza, Utah, west of Rangely. From that wintering area sheep could be summered in either the Colorado mountains back toward Meeker or in the high Uintah Mountains north of Vernal. Paco also spent a considerable number of summers herding in the Gunnison Basin east of Montrose and in the mountains of the San Juan Triangle about Ouray, Silverton, and Lake City, Colorado. These sheep were apparently wintered to the west in the canyon country of western Montrose County, Colorado, or just across the line in eastern Utah. Most of a herder's time was spent alone on either winter or summer ranges.

The most intense activity occurred at lambing time in the spring and in the gathering and sorting time of the fall. These activities required the presence of more people than just the herder. These two occasions were what typically brought the herder into the most contact with his bosses and other herders. The rest of the time, he was usually alone with only his dogs and the sheep for company. He might occasionally see his boss or a camp tender who resupplied him with groceries every

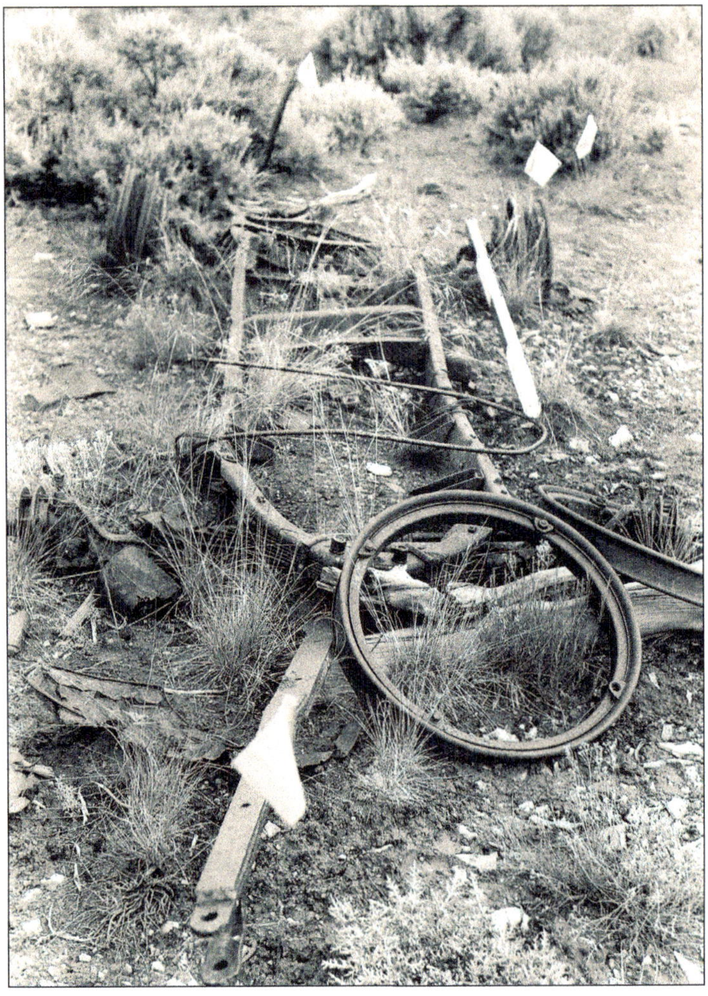

Figure 16: Disasters did happen out on the sheep ranges as this photo of a burned out homemade sheep wagon attests to. This wagon had been built on an old car or truck frame and burned to its chassis, probably in the 1930s or 40s. This site was near Rangely, Colorado (author's collection).

few weeks. If he was close to a decent road and he was not too far away, his family might even be able to make a quick visit to the herder's camp.

Before the sheep were taken to the winter range, the herder might be able to slip away for a few days or weeks at home with his family if all the flock's needs had been covered, and he could find someone to take over for him. But sheep could never be left unattended. At Christmas and other holidays, or even major family events--even funerals--there was usually no opportunity to get home or partake in any revelry. One day was pretty much like another. Before the advent of modern communications, a herder might not learn of a death or birth in the family for weeks. In Paco's case, he could not even get home until after Ophelia had died after a long bout with cancer.

Winter camps were often quite remote and accessible only by roads that were likely impassable in bad weather. Summer camps were, likewise, typically deep in the mountains far away from roads. A herder was on his own, particularly in the days before CB radios and cell phones when there was no way to call for help in an emergency. Paco once confided to me that, although he learned to accept the loneliness, his biggest fear was getting hurt or sick and dying alone out in the hills. In such an event, he was afraid it might be some time before his body was found. If he were to die away from his camp, he feared his remains might fall victim to wild animals.

Surely the annals of sheepherders must have many accounts of real tragedies such as Paco feared. Although Paco maintained relatively good health into old age, in 1966 he contracted Rocky Mountain Spotted Fever while out with the sheep. He became so ill that he had to be hospitalized in Craig, Colorado, and then missed the entire 1966 herding season. During his long recovery, he stayed with his daughter, Alice, in Fruita. While this book was being completed, a Peruvian sheepherder had a close call working in Utah's La Sal Mountains. The man was just resting under a tree when a bull elk suddenly appeared and attacked him. Although gored through one lung, knocked unconscious, and otherwise badly banged up, the man managed to walk five miles to seek help from another herder (Associated Press 2013).

There was an established basic daily routine for herders, but each one had his own way of structuring his day and handling his sheep. Camp had to be relocated periodically depending on grazing conditions. If the sheep were moving any distance, the herder's camp also had to move. Back when tents and pack animals were still in use, a camp tender was employed to pack up and move the camp for the herder. Paco once outlined his own typical daily schedule while out on the range with his charges. This usually did not vary much no matter the time of year or the weather. He followed a similar routine for months on end.

At 4:00 AM he got up and built a fire in his little sheepherder's stove in the tent or sheep wagon. The routine was pretty much the same no matter what kind of shelter he lived in. He made fresh coffee, and his breakfast consisted simply of coffee and cheese.

By 5:00 AM Paco was on the move. The sheep were starting to get up from their bedding ground and spreading out in search of food. The herder had to be attentive at this time to keep the sheep together and begin drifting them in the best direction for grazing. If coyotes were in the area, they tended to strike the sheep while they were still spread out after breaking up from their tight bedding formation.

Paco had to be especially alert in the early mornings because of this danger. The key to successful grazing of the sheep was to ease them along slowly without rushing them. If moved

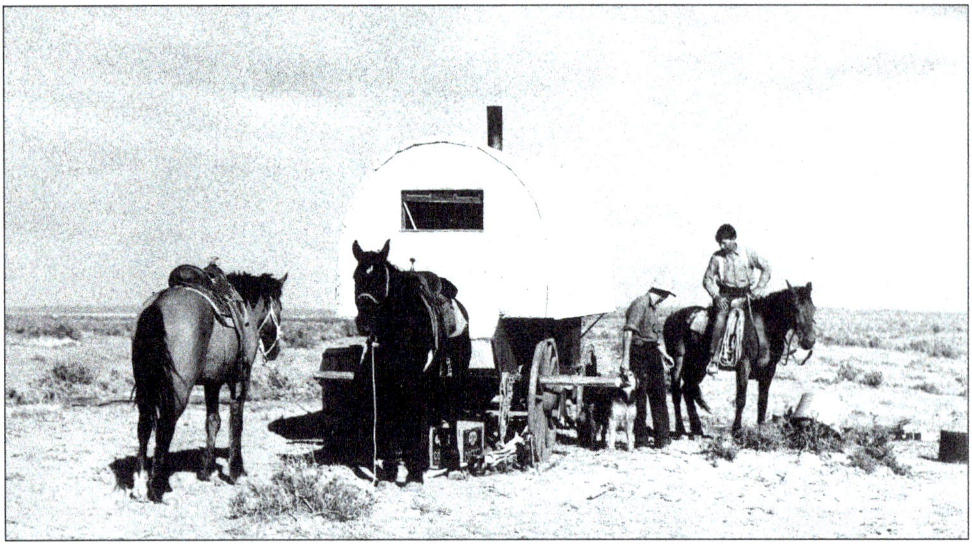

Figure 17: An undated though likely early twentieth-century photo of sheepherders, their dogs, and sheep wagon home somewhere in the western U.S. (courtesy Western History Department, Denver Public Library No. F32361).

too fast, they became strung out and badly scattered. It was then difficult to bring them all together again. He also had to be careful not to force them or let them drift into heavy timber. There were often lots of deadfalls and snags that could harm the sheep and even puncture the bellies of the lambs.

Between about 9:00 and 10:00 AM it was time to water the sheep. The herder had to select a safe approach to the water hole that was not too steep or rocky. After drinking, the sheep tended to lie in the shade, out of the heat close to the water source, for a few hours. The herder encouraged them to do this because it allowed him time to leave them and return to camp. He used this time to chop firewood, haul water, make bread, and eat. He commonly baked his bread in a Dutch oven covered with coals. Once these duties were tended to, he might try to get in a short nap. This was also a good time to carve or pursue other hobbies to help relieve the boredom. In addition to carving, herders have been known to pursue all kinds of diversions, including looking for Indian relics, doing a lot of reading, or erecting stone cairns known as "stone Johnnies."

By 3:00 PM Paco was back at the waterhole with the sheep. They were starting to move again looking for better grazing. He let them do so but could not allow them to move too far as he did not want them to trample over good grass. He wanted them to eat their way along slowly. Again he had to remain attentive as the sheep at this time began to scatter widely if the herder and his dogs did not hold them in check.

At about 5:00 PM, Paco headed the flock slowly toward a salt ground that was to be their bedding ground for the night. The sheep required a different bedding ground every day. He bunched the sheep up as tightly as possible, and they continued to bunch up even more tightly on their own. This was another time coyotes might prey on the flock. By about 7:00 PM the sheep were usually tightly bunched up and bedded down. Once he was sure they were well settled, Paco could walk or ride his horse back to his camp and have his evening meal. He then read and listened to his portable radio before going to sleep. Paco was particularly fond of murder mysteries. He also took this time to plan the flock's movements for the next day.

In his early years as a herder, Paco, like many others, was not provided with a horse and had to perform all of his duties on foot. In his later years as a herder, Paco had a horse and the added luxury of living in a relatively well-appointed, modern sheep wagon. This was a far cry from his early days as a herder when his home was a ten by twelve foot square white canvas wall tent. Such tents are still appropriately known as "sheepherders" tents. When tents were still in use, the employer hired a camp mover or tender. This man's responsibility was to move the camp as necessary to keep it close to the grazing grounds. This might be every few days or weeks depending on conditions. One mover might tend to more than one herder's camp.

A big sheep outfit might have multiple herders out in the hills at any one time, and the tender had to keep moving the camps as well as resupplying them with groceries and checking on the herders. The camp tender typically utilized a string of three or four pack mules for making his rounds to resupply and move the camps. He also had other duties. He packed salt for the sheep, and if the herder was working above timberline, he even had to pack in firewood

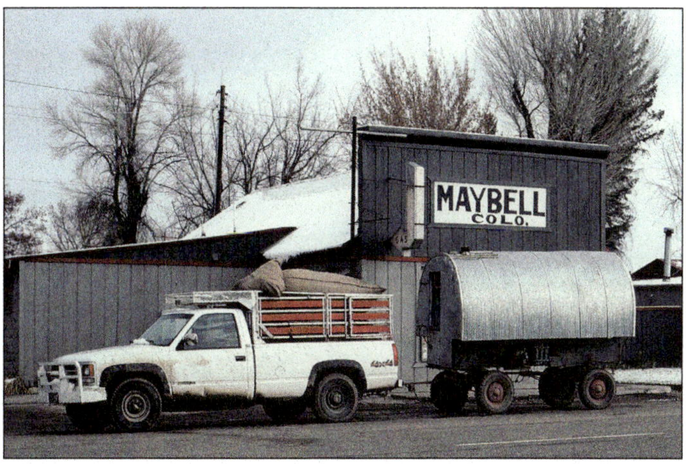

Figure 18: A sheep camp on the move in the spring of 1991 at Maybell in northwestern Colorado. The pickup contains bags of wool gathered during the spring shearing. This view shows how the herder's sheep wagon home had evolved from the old horse drawn and wooden wheeled ones (author's collection).

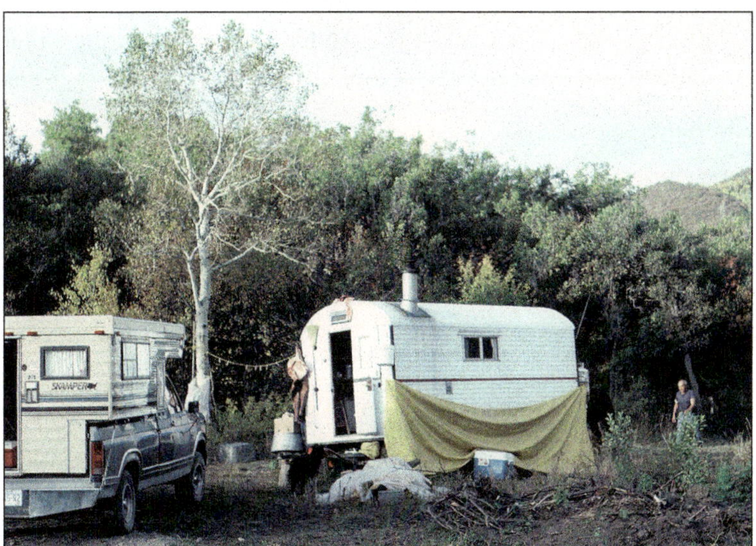

Figure 19: The camp of Presciliano Chacon on the summer range at Yellow Jacket near Meeker, Colorado, in October 1992 before the sheep had been gathered in preparation for moving them to winter range. This view shows how the sheep wagons have continued to evolve (author's collection.)

for him. Although the camp tender rode a horse, the herders often did not have one in the early days. Riding "Shanks mares" was the common lot of many of the early day herders.

Back when there were still camp tenders, the herder enjoyed their company at times, and this helped to relieve the loneliness. If the tender was caught up with his other duties, he sometimes stayed with the herder for a few days and helped him stockpile firewood, haul salt, and do whatever else needed to be done. These stays seldom lasted more than four or five days as the man's duties to other herders once again had to be taken care of. The tenders usually carried their own small, cone-shaped tent that was suspended from its peak by a tripod of exterior poles. These were known as "sheepherder's tepees" and were at times used by the herders themselves in the old days (Weidel 2001:41).

A good employer usually tried to visit the camps of each of his herders periodically. If the camp was not too remote, he might appear once a week or so. Some of the smaller employers were herders themselves and thus directly in charge of some of their own sheep. When the boss appeared, the herder provided him with a list of groceries and other items that might be needed. If there was no camp tender the employer personally brought in the groceries on his next visit. It was, however, not uncommon for the boss to have a host of other obligations that were more urgent than checking on his herders. There were times when the herder's supplies ran so low that he was on meager rations for some uncomfortable periods.

Paco recalled at least two bosses he had over the years who were notoriously stingy with food. The employers have always been responsible for the board of their herders. Paco recalls making out grocery and miscellaneous supply lists and giving them to these employers. When the goods finally arrived, however, the stingy employers had "forgotten" to get a few items or simply decided on their own that Paco really did not need as much of a particular food as he thought he did. If the employer was intending to stay with Paco for a few days, as he sometimes did, the grocery list was usually completely fulfilled and perhaps even extended a bit.

Typical fare for Paco included lots of coffee, sugar, cheese, carrots, potatoes, onions, a variety of canned fruits, and other vegetables and meat as might be requested and available. Flour, salt, yeast, and crackers were also basics. There are reports of herders taking some wild game, such as antelope, deer, or elk, and it was not unusual for a herder to eat one of his

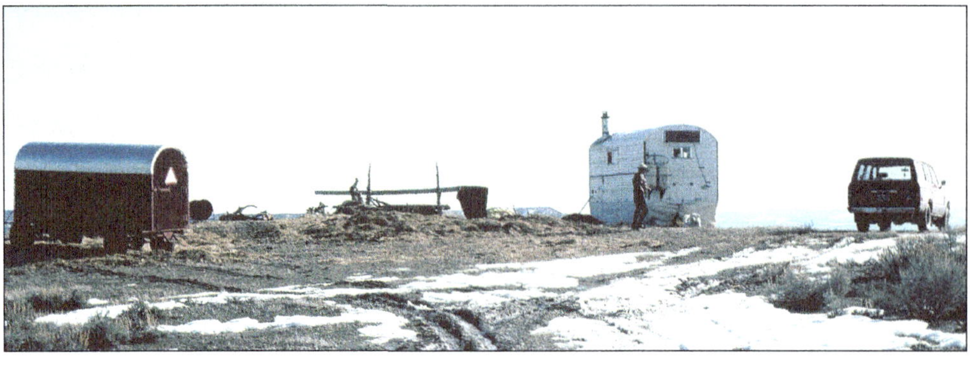

Figure 20: The camp of Presciliano Chacon on the winter range south of Bonanza in northeastern Utah in March 1992 before the sheep had been gathered for lambing, shearing, and moving to summer range. This view shows how the complete, two-unit, modern herder's hookup has evolved from the old style shown in Figure 15 (author's collection.)

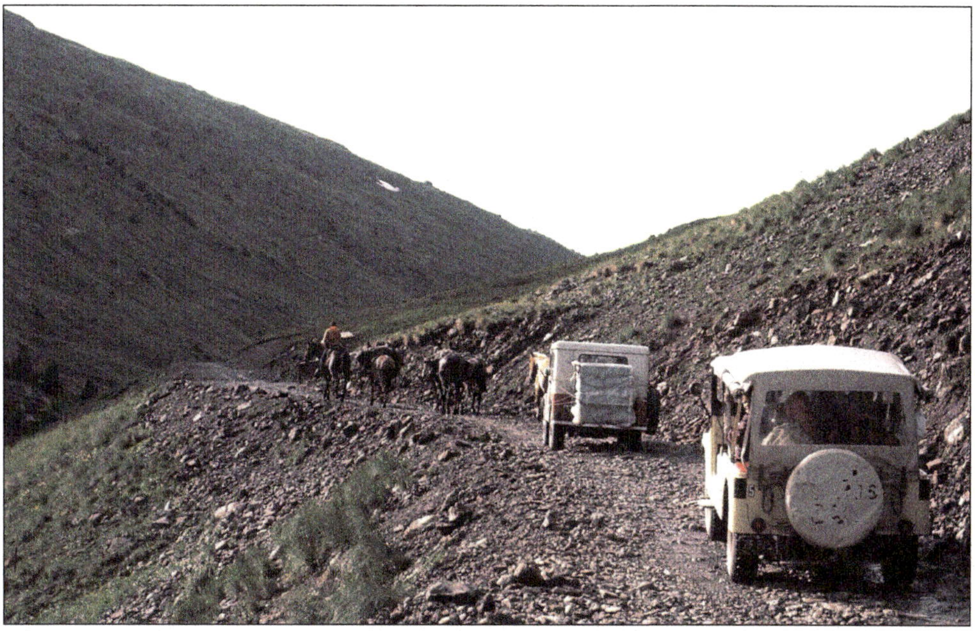

Figure 21: A 1970 view of a sheep outfit's old-style camp tender moving or resupplying a sheepherder's summer camp by pack mules and holding up a tourist jeep(s) near Mineral Point in the San Juan Mountains above timberline near Silverton, Colorado. The red and white jeep seems to be carrying cargo and may be part of the sheep outfit (author's collection.)

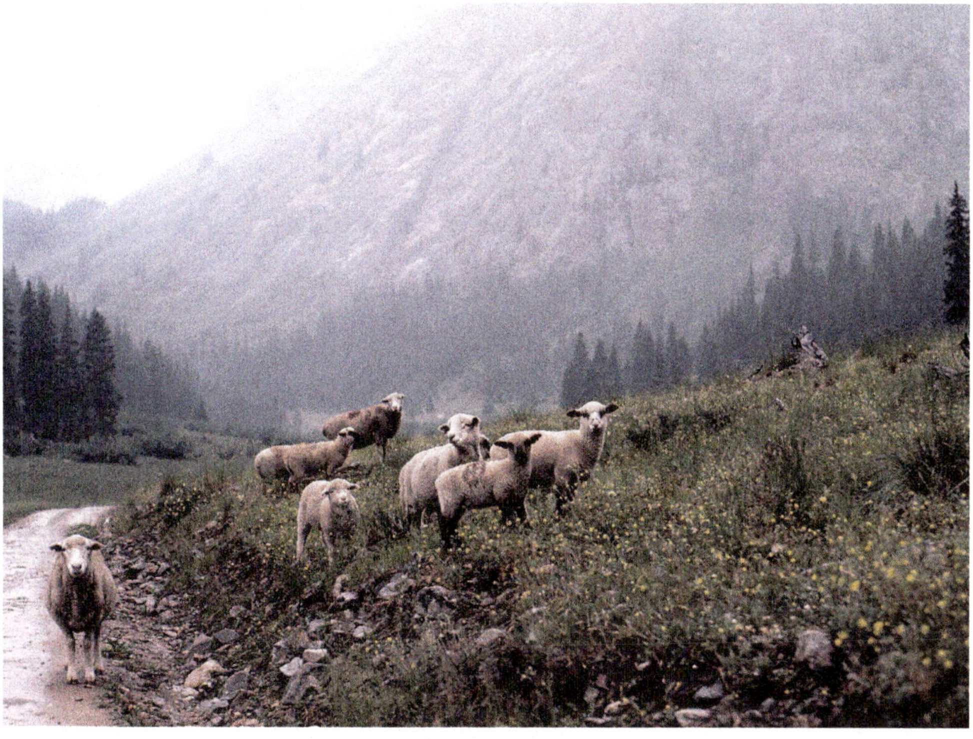

Figure 22: Sheep, including young market lambs, on a rainy day in 1970 in the summer pasture high in the San Juan Mountains above Silverton, Colorado (author's collection).

sheep, particular if it was injured or suffered an accidental death. Preserving fresh meat in the mountains without refrigeration required some ingenuity. Paco preferred to wrap the meat in a clean meat sack and tie a rope around the sack. The sack was then suspended in the cool shade high in the trees. This old method is also a still useful means of protecting groceries from bears. It thus served two purposes. The sack itself might be wrapped in multiple layers of canvas to further insulate it. Meat could be kept fresh by this method for several days in the summer heat so long as it was kept shaded.

Fresh vegetables such as carrots, potatoes, and onions were preserved by burying them in a shallow trench in the ground in a shaded area. With the advent of insulated ice chests, these old ways of preserving fresh food fell to the wayside. Even though ice might not be available, the chests could be kept cool by placing gallon jugs of cold spring water in them and keeping them shaded.

When Paco was still using a wall tent he arranged it in a fashion that was typical of most herders. Firewood was stored to the left against the wall just inside of the doorway against the canvas. This wood had been cut to the correct length to fit the particular size of sheepherder's

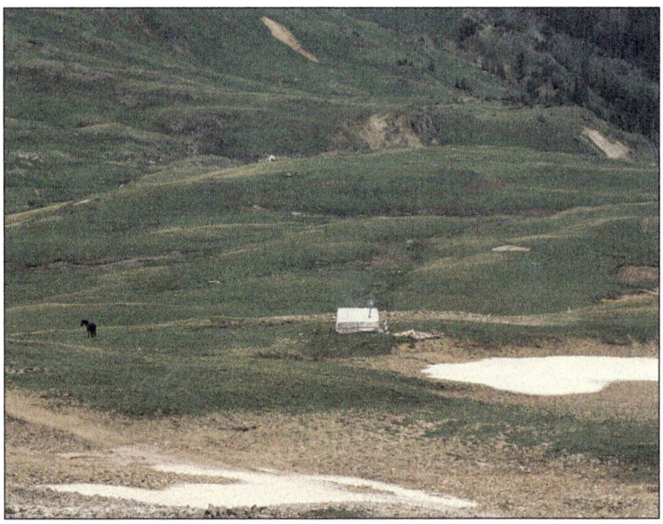

Figure 23: A 1970 view of an old-style, traditional sheepherder's tent camp in the high alpine tundra of the San Juan Mountains near Mineral Point above Silverton, Colorado. Although someone somewhere out on the sheep ranges may still be using tents, this kind of herder's camp, which was once common, has now almost disappeared from the Colorado landscape. Early in his career, Paco Chacon herded in this area and almost certainly utilized a tent like this (author's collection).

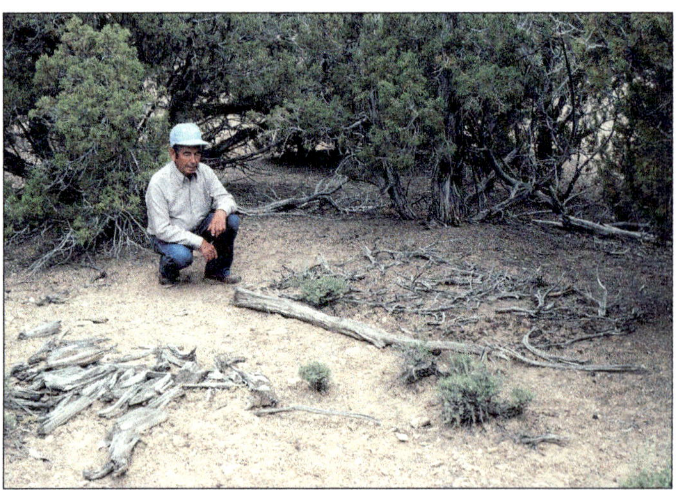

Figure 24: Paco Chacon examining the site of an early twentieth-century, typical sheepherder's tent site in old winter range south of Rangely, Colorado. Note the firewood in the foreground is all cut to the same length to fit the specific stove used by this herder. In the rear of the tent, the herder had constructed a bed of juniper tree boughs as now only marked by dried up sticks and twigs. View of 1991(author's collection).

stove that was being used. This stove was small and made of sheet metal. It was placed adjacent to the firewood with its pipe passing through a metal flashing or collar riveted into the sloping roof of the tent. Stones placed on the tent floor often supported the stove. A wooden grub box or camp cupboard was placed against the tent wall opposite of the stove and firewood. That was the kitchen where groceries and cooking equipment were kept. I have recorded a great many old herder's tent sites, and the archaeological evidence is consistently as Paco described it (Figure 24). The location of the stove is commonly marked by heat-reddened stones that were used to support it.

The herder's bed was located against the back of the tent (Figure 24). In the old days, freshly cut tree boughs were used as a mattress. In his early years, Paco's bed consisted of a cotton-filled mattress on a canvas laid on the ground. His bedding then was blankets. Even when sleeping bags became available, Paco preferred to use blankets because of the freedom of movement they allowed. Paco eventually began using a folding, military-style cot with an air or foam mattress. There were, of course, no toilets or bathing facilities other than a washtub and a shovel with a roll of "tp" stuck on the handle or a primitive privy. Although there may be a portable toilet somewhere out there in "sheepherder land," these basics of personal sanitation have changed little through the centuries.

Paco gave up tent living in 1959 when he received his first sheep wagon from Louis Livingston of Craig, Colorado. By the 1970s, Occupational Safety and Health Administration (OSHA) standards were being implemented in the sheepherding industry, and the old style wagons were being phased out. By that time the charismatic, wooden sheep wagons had become collector's pieces (Figures 15, 17) and were being replaced with modern, highway-worthy trailers with metal skins, rubber tires, and even solar panels (Figures 18, 19, 20). Today these are the norm for most herders.

The interior layout of sheep wagons has remained basically the same as that used in the old tents with the bed in the back and the stove and kitchen areas on either side just inside the door. Village blacksmiths or the larger sheep ranchers originally built sheep wagons locally. It is not unusual to see a variety of "homemade" sheep wagons commonly (Figure 18) built by smaller operators. Modern sheep wagons (Figures 19, 20) were then for a time manufactured commercially by companies specializing in them. Today there is only one manufacturer still making sheep wagons. This is Wilson Camps, Inc. of Midway, Utah (Wilson Camps 2013).

> *A sheepherder's wagon is small,*
> *Just a bunk and a stove and a wall.*
> *He can cook and can sleep*
> *And can tend to his sheep,*
> *But overnight guests? Not at all.*
> (Peterson and Rhodes1985:17)

One of the primary duties of a herder is to protect the sheep from the ever-present predators such as coyotes, bears, and mountain lions. Paco (Figure 25) was always attentive to this aspect of his responsibilities and took pride in the comparatively low losses his flocks suffered from such predators. He recalled that bears were a serious problem on the Uncompahgre Plateau in the summer of 1938. That summer he personally trapped and killed eleven marauding adult bears and his older brother, Jose, who was herding nearby with his own flock, killed fourteen. The secret to defeating coyotes was to keep the sheep bunched up. In this Paco was successful, and he suffered low losses to coyotes.

One winter Paco was herding in the canyon country along the Green River on what is now the Northern Ute Indian Reservation in Utah. A mountain lion was continually stalking his flock and had taken many lambs. Paco tried every trick he knew to kill the lion but could not get it. By the time it had taken a dozen or more lambs, Paco felt obliged to call in the government trapper who finally killed the destructive lion. In celebration of the event, Paco carved an image on a cliff face. This showed a young lamb being taken down by a mountain lion. Paco, and his brother, Press, always stressed that this was probably the finest work Paco had ever carved on a rocky cliff face. Unfortunately, the Northern Ute Tribe will not grant me or anyone else permission to go into Tabyago Canyon to find and photograph this particular carving. Paco left other carvings in that area, but apparently no one other than Paco and Press (Figure 26) and perhaps a few others has ever seen these works. It is most unfortunate that they could not be included in this volume.

Typically, a herder's only steady companionship was from his dogs, horse, and the sheep themselves. Paco trained his own dogs by pairing the young ones with older experienced ones. The older dogs did most of the training. Visits from family were rare, but the herder might occasionally see his boss or the camp tender.

To be a successful and happy herder a man must be content with his life and himself. One of the classic and most insightful studies of sheepherders was done by Michael Mathers (1975:7). In that excellent photojournalistic work, the author asks what kind of men sheepherders are and why they choose a lifestyle where they have to live alone. Mathers concluded that they were a unique breed of men. One sheep rancher believed they had to be crazy to live that way. As the highly literary herder Archer Gilfillan (1929:3) stated, Mathers thought that if they had not been crazy to start with, they certainly became crazy in time. Some were just plain cranky and not content with their lot as exemplified in the 1920s aspen carvings of the Mexican herder Frank Rodriquez who stated:

> *Who reads this will kiss my ass.*
> *It will be some shit sheepherder who*
> *loves to know that it is none of*
> *his business.*
> (McGonagle 1990:41)

Rodriguez also stated: "... *who is a sheepherder is not a man, is a wretch, let them be eaten by the coyote* (McGonagle 1990:41)."

But after spending much time with herders, Mathers came to believe that any such craziness actually had a large vein of sanity woven through it that most non-herders do not have. He also believed they were truly peaceful people who were strong and self-reliant as well as gentle and compassionate. Alone on the range, a herder becomes a part of the flow of nature without hurry or impatience. He nurtures and gives to the living things about him. In light of my experience with sheepherders, particularly Paco Chacon, Mathers's impressions seem accurate for the good and successful ones.

These are the qualities that those who fail at the profession do not possess. There are, however, two sides to a herder's personality. One is when he is alone, and the other when he congregates with other people, particularly with other herders. When he was sober, Paco displayed these traits and was a gentle and peaceful man. When drinking he could display a rougher edge. Family members often described his brother, Press, as just plain mean, particularly so when he was drunk.

Although eccentricities are present among herders when alone on the range, they commonly, although not universally, display a pattern of gregarious irresponsibility when off the range and gathering with their fellow herders for a short period of communal insanity. This is where the notion of the "crazy sheepherder" on a "tear" or binge of drunkenness comes into play. Such tears typically occurred in the fall at the end of the summer season before the sheep were taken to the winter range.

Like liquor, after so long alone, sex ranked high on the list of a herder's "things to do" during these brief returns to civilization. Although chronicles of drunkenness are easy to find, the women in sheepherders' lives have left few narratives of their experiences. One woman who dated and knew sheepherders well gave her own graphic personal perspective on the issue. Her comments are worth the reader's time and may be found in Mathers (1975:89).

Paco never spoke to me about his love life. But in terms of his penchant for liquor, Paco was certainly true to the calling of the crazy sheepherders at this time in the annual cycle. He might get home to Fruita for up to a month with his family or a girlfriend. Both before and after his separation from Ophelia, he brought money and groceries to her and their children. Still, he and his herder buddies went on a tear and stayed drunk for much of this time. Though he provided for his wife and children, he never asked what they needed. He simply went to

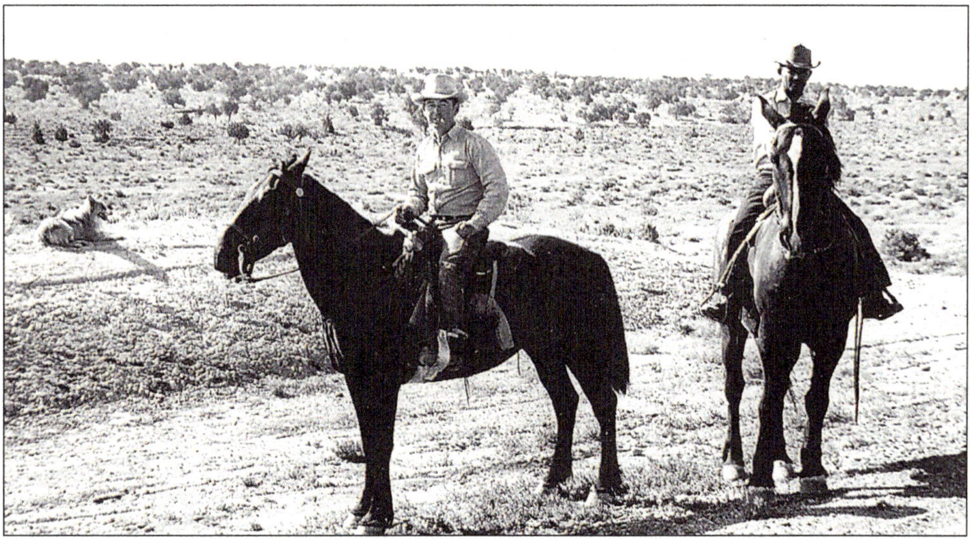

Figure 25: Paco Chacon and his boss Mike Theos herding sheep. This view was most likely taken in the 1970s in the Shavetail Basin winter range south of Rangely, Colorado (courtesy of Alice Montano, Grand Junction, Colorado).

the store and bought lots of groceries. This was a continual sore point with Ophelia. He also brought home lamb and wild game such as elk.

With much of his money spent, Paco would, once again, head out to winter range with his pockets turned out. The next months alone allowed him to regain his composure, begin to rebuild his stash, and provide time for his body to repair any damage he had done to his liver and brain cells. The herder's seasonal schedule then began anew. With almost clockwork precision, the annual ritual cycle played out, and the seasons turned much the same year after year.

Eventually a herder's body wore out, and he had no choice but to quit the hard life. Some men like Paco had family nearby and the good fortune to have Social Security and Medicaid to support and care for them in old age. I can only wonder how the old-time herders, who usually ended up poor men, managed their retirement in the days before Social Security. The great poet, Walt Whitman (1964:53), once stated in his poem, The Death of the Hired Man, that "home is the place when you have to go there, they have to take you in."

In the old days, an aged herder with a family to take him in must have felt fortunate. Back then, some thoughtful sheep ranchers are known to have allowed good, retired long-term employees to continue to live out their lives on the ranches. They might give them food and

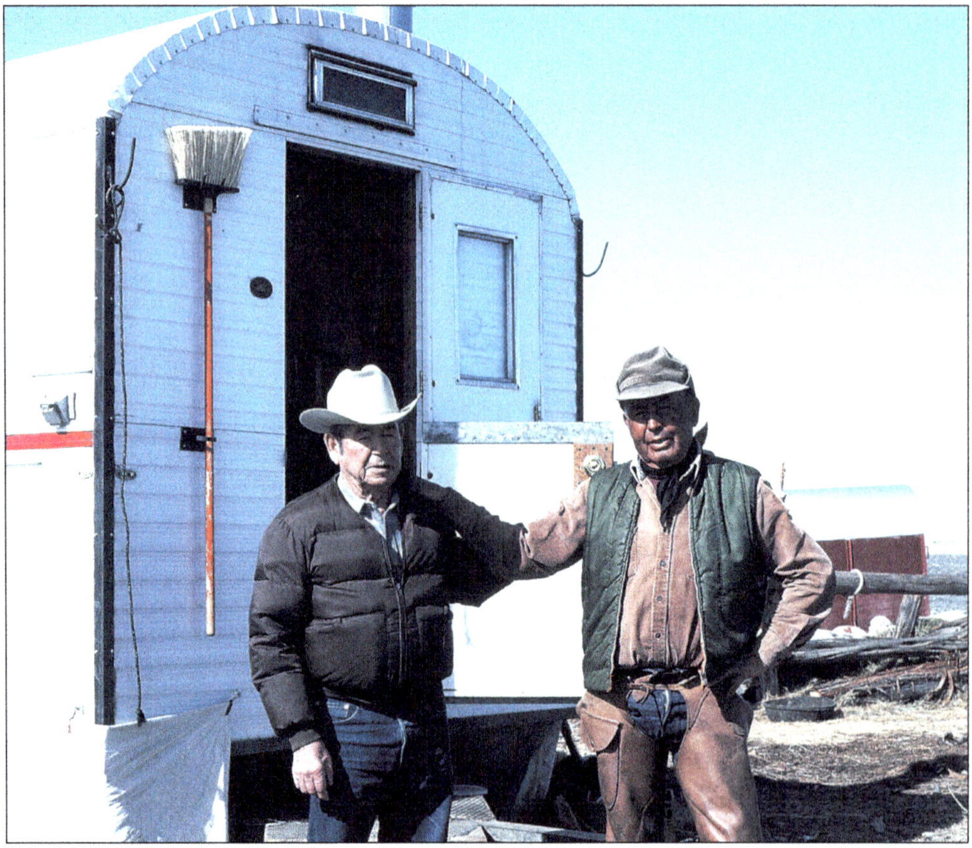

Figure 26: A retired Pacomio Chacon (left) and his younger brother, Presciliano, at his winter sheep camp south of Bonanza, Utah, in March 1992 (author's collection).

a small allowance in exchange for doing those small chores they were still capable of carrying out. The plight of those with no one to take them in was certainly problematic.

Paco had his Social Security and a small military pension. He was also fortunate and astute enough to obtain a little piece of land with a small trailer home in Fruita near his family. In his later years, he shared this home with his brother, Press, after he too retired from herding. Prior to his retirement, Paco commonly stayed with his daughter, Alice, at her home in Fruita when he was not with the sheep. She worked to prevent him from drinking, and he would tell her that she was a "mean woman" for doing so!

Now all his flock are bedded down
where wind
Has bared the high hill slope like
a big hand
pressed warm upon the snow,
and stars have thinned
The dark with blue-white jets of
ancient fire
And made the covered wagon
silver-clean
As some great sail lifting upon
the lean
And lonely wave of land that is a
ridge.
The herder pauses by the wagon
door
And hears wind pluck the canvas
roof and sees
The hills sharp-etched against the
neighboring sky.
Then shadowlike the sheep dog
rubs his knees
As though the master called him
close to share
Some secret of the night, some
small lost sound
Singing between the vast sky
and the ground.

(Sheepherder in Winter [Curry 1951])

Chapter 6

PACOMIO THE ARTIST

Sheepherders are known to have often been quite creative in the ways they found to prevent or relieve their loneliness and boredom. Some read a lot, while others whittled, collected Indian relics, or stacked rocks into free-standing monuments (Figure 27) known as "stone Johnnies" or "stone boys" (Allen 2008; Harris 2010, 2011; Mathers 1975:9, 56). Many probably played a lot of solitaire, and there were even a few who were writers and poets. One old herder apparently even carved a big elaborate castle from solid rock near Willow Creek in the Gunnison Drainage of western Colorado (Figure 28). Those who could not find healthy ways to entertain themselves were most likely the ones that did go mad and become the truly "crazy sheepherders" of traditional Western American folklore.

Among all of the pastimes, however, carving on the aspen trees of the summer ranges was almost universal. Few herders could completely pass up the opportunity to quickly leave a note, carving, or just their moniker on an aspen tree at some point in their careers. Some were very prolific and left many such carvings. A very few, like Paco Chacon, also left such carvings on rocky cliff faces.

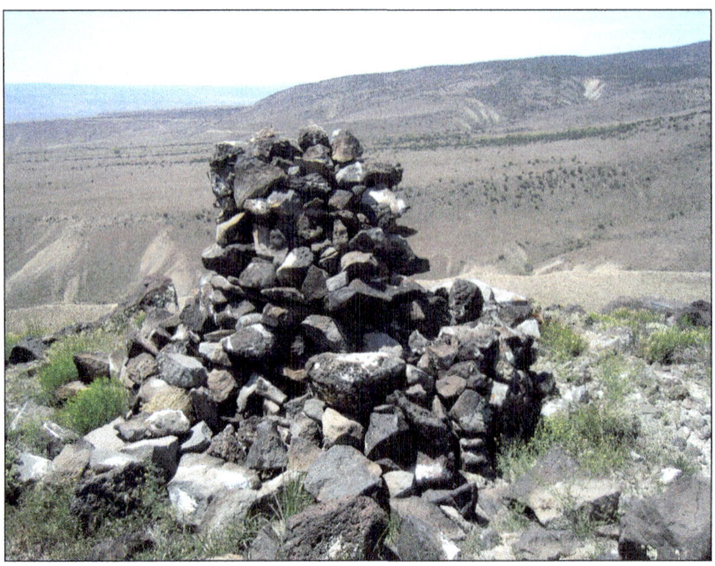

Figure 27: A stone cairn that is probably a sheepherder's monument known as a "stone boy" or "stone Johnnie" from old sheepherding winter range north of Delta, Colorado (courtesy of Bill Harris, Montrose [Harris 2012]).

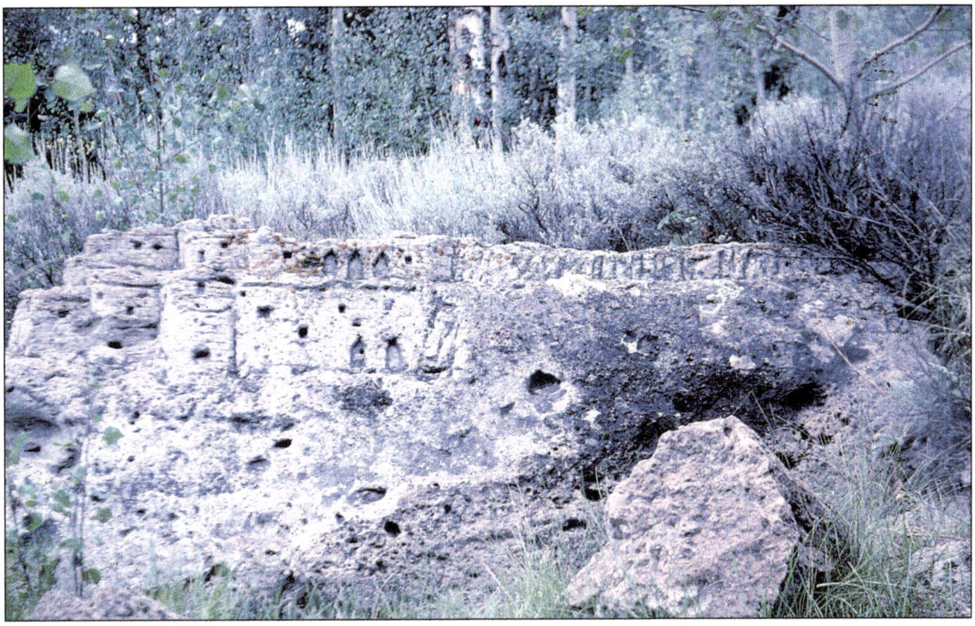

Figure 28: A Moorish-style castle believed to have been carved by a sheepherder long prior to about 1950. This was from the Willow Creek area between Blue Mesa Reservoir and Lake City, Colorado. It has been moved and is now near the recreation center at the Blue Mesa Subdivision in Gunnison County, Colorado. If actually the work of a sheepherder, it was likely carved over the course of multiple short visits to this specific location since it would seem to have taken considerable time to carve (1960s photo courtesy of Monte and Shirley Sanburg, Montrose, Colorado).

An old poem states quite well some of the things that could lead sheepherders to pastimes such as building stone monuments.

Silent pile upon the hilltop,
With vastness all around;
Landmark of the lonesome shepherd,
Sit 'mid silence most profound.
Guide for men and flock returning,
O'er the hills near daylight's end.
Man who has no other helper,
Than his faithful canine friend.
Monument of patient labor,
Builded slowly stone on stone;
Built to while away the hours,
Hours he spends there all alone.
Who can tell what thoughts come o'er him,
As he worked there all alone!
Some past tie he may have severed,
With the placing of each stone.
Each stone may be a hope departed,

Or plan, that now lies buried deep,
While memories crowd in upon him,
In close array like flocks of sheep.
This pile of rocks may not be classic,
It may not be a work of art;
But it may be the sign of longing
In some poor sheperd's aching heart.

(Untitled poem by T.S. Parsons [1920],
referenced in Shroup [2008]).

The reasons why Paco Chacon carved on sandstone cliffs and white-skinned aspen trees are quite simple. First, as a sheepherder, he had lots of time to do so. Once his sheep were appropriately pastured or bedded, other than for his routine camp chores, there were few other activities to compete for his time. The cliffs and trees were readily available to him. Paco also had enough talent and appreciation for art to selectively choose to make use of the available time and canvases. He also had a creative and fertile mind that allowed him to envision his subjects with considerable clarity. He then truly mastered the techniques necessary to transmit very accurately and eloquently his memory visions of the subjects to the tree bark or cliff faces.

Paco labored to perfect his art, and through the years it evolved in a pattern that can in part be traced through the examples in his portfolio. His efforts to perfect his art most likely commenced as a small boy doodling in his school tablet. As a teenager he worked to improve his handwriting and enhance his already well-developed signature. He did not want to be thought of as just an uneducated herder who could barely write his name. Through trial and error he experimented with new techniques for carving his signature and pictures on cliffs and aspens. He sometimes found it a challenge to do his best work in these mediums.

Paco was admittedly proud of his work ethic and sheepherding skills. He was good at protecting his charges and through fat market lambs brought profits to his employers (Figure 29). He could readily talk fondly at great length of sheep and the herding life way and his experiences within it. He loved to mimic his bosses and find humor in other people's foibles, sometimes using only facial expressions with few or no words. In this regard he was quite a mime and a very "Chaplin-Like" figure.

His artistic sensitivities were very clearly coupled with his powers of personal observation of people and nature and his ability to mimic them. He could be quite animated in his verbal accountings. In his gestures and facial expressions, he relied upon the same whimsical and chimerical qualities so evident in his renderings of the people illustrated in his portfolio. One such tale bears repeating here.

Paco once worked for a Colorado employer who had a notably profitable year, in very large part due to Paco's efforts in bringing in a fine crop of fat lambs. After the lambs had been sold at good prices, his employer made a trip to Las Vegas where he lost over $10,000 gambling. He did not even have enough money left to rent a motel room on his way home to Meeker. According to Paco the man had to sleep in a ditch alongside the road.

Paco was furious over this incident and took it as a personal insult to his efforts to produce profits for this employer. In retelling the story to me on various occasions, he always became extremely agitated and animated. He made faces and used his hands to elaborate his points. His English was typically spoken softly and was strongly inflected with the accent of his native Spanish. Although he was particularly fond of this particular employer, in a sarcastic, somewhat childish, and soft but highly mocking, tone he explained what he had said, or hoped to have said, to him as closely paraphrased below.

> *You dum' sonaofabitch! Wit' dese hands Paco*
> *wor' so hard for a year to brin' you fa'*
> *lambs wit' silver dollars stack on*
> *der backs. Wha' a dum' f... you are!*
> *I brin' you de lambs an' you go a la Vegas*
> *an' get all drun' up and gamble all my*
> *har' work away! You no even save enough*
> *money to come home on and have sleep in*
> *de ditch. You have no respect for Paco!*
>
> (Pacomio Chacon to Steven Baker on multiple occasions
> in Fruita, Colorado, ca. 1992 and later).

Still, his art is what most readily made Paco smile and become animated. He was not in any way a boastful or bragging type of man and really enjoyed drawing and carving. He was also gracious when discussing the carvings of other herders, although he always knew they could not hold a candle to his. He admitted that he found carving to be relaxing. In addition to helping him pass the time, his ultimate goal in producing the works was as a way of communicating with those few people who might be out tromping around the woods and canyons. As Paco put it, he simply wanted to leave something behind to make people "smile and be happy" and recognize that "Paco had been here." They were his calling cards.

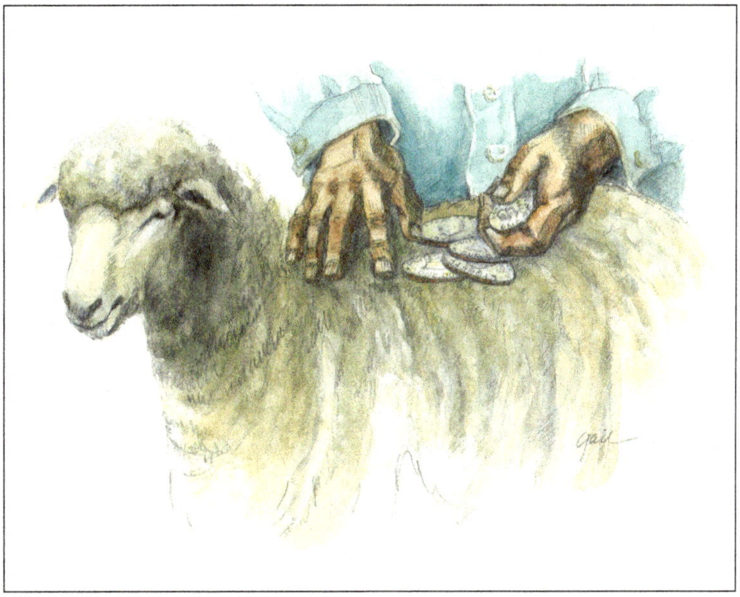

Figure 29: Paco always envisioned his efforts on behalf of his employers as through his hands he brought them silver dollars on the pelts of fat lambs (watercolor painting by Gail Carroll Sargent, Centuries Research, Inc., Montrose, Colorado).

But as I understood Paco, there was more to it than that. Despite the fact that Paco was a self-reliant and successful man who had come to accept his lot in life as a herder, he was still somewhat uncomfortable away from the sheepherding world and outside of his limited comfort range in modern society. He often said he was "just a sheepherder" and explicitly understood that he had comparatively little education or social, political, or economic status. He acknowledged that he occasionally drank way too much. Yet, there was one place where he enjoyed considerable status among sheepherders, his employers, and the establishment and that was as an artist, and he was not too humble to bask in it.

He had seen a great many of the carvings other herders left behind among the aspens (Figures 30-33). He was certainly no fool and recognized his abilities and how they ranked among these other carvers. He was proud to hear the many compliments made to him regarding his work. Although he was not an ego-centered individual, all people have to have their ego stroked now and then. Paco's came from other people's recognition of his abilities as both an honest and hard-working herder and an accomplished folk artist.

Paco was known to commemorate special events in his carvings, such as his cliff face carving in recognition of Christmas Day 1976 with "Paco's Nacimiento" (Navity scene, PRA 23 in his rock art portfolio in Chapter 8) or his still unrecorded rock art carving that he completed when the marauding mountain lion had finally been killed on the winter range in Utah. He also carved his dogs, horses, and wildlife that he encountered. Although he drew pictures on paper of sheepherding activities, he seldom if ever made carvings related to them.

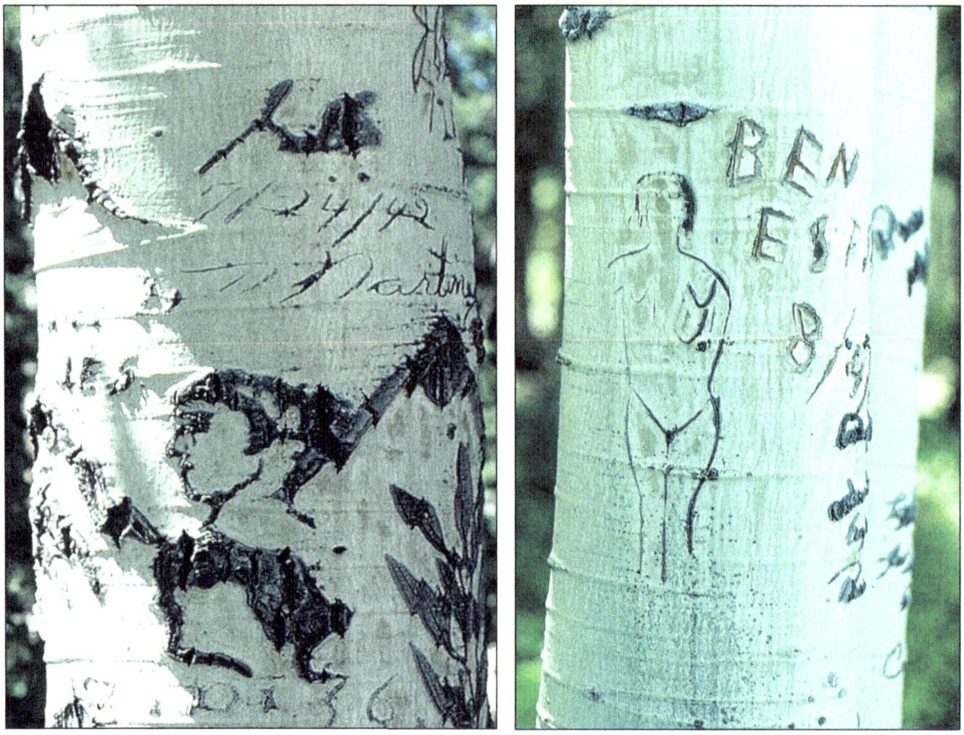

Figure 30: Examples of typical sheepherders' aspen carvings of women from the Lone Cone vicinity of the Uncompahgre Plateau near Norwood in west-central Colorado (views of 1976 from author's collection).

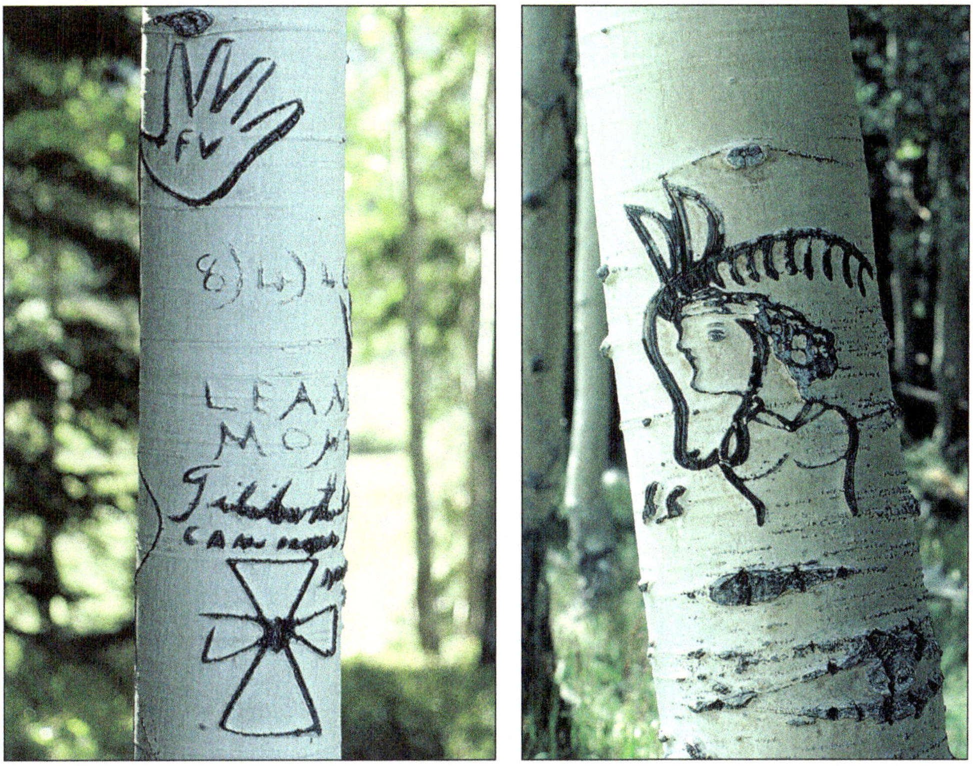

Figure 31: Examples of typical sheepherders' aspen carvings from the Lone Cone vicinity of the Uncompahgre Plateau near Norwood in west-central Colorado (views of 1976 from author's collection).

Most of Paco's carvings represent imaginary people. His women in particular, with the exception of his depiction of Marilyn Monroe (Figure 1 and PAP 17 in his aspen portfolio in Chapter 7), were only imaginary. For a long time, I suspected that some represented women he had known. Paco always adamantly insisted that they did not, and I ultimately came to believe him. On one occasion I asked Paco why he drew beautiful, sexy women. He responded quite honestly by stating, "Ay yay yay, that is all I could think of when I was out there!"

Carvings on trees are formally known as "arborglyphs" or "dendroglyphs." People the world over have left all manner of carvings on trees over thousands of years. Aspen carvings are a form of what archaeologists refer to as "culturally modified" trees. Many types of trees have been modified for a wide variety of purposes. The "blazing" of a tree with an ax to mark a trail is an old practice, and remnants of such blazing can still occasionally be seen in the forests of the American West. One can surf the internet and find endless examples of culturally modified trees.

Virtually everywhere sheepherders lived and worked among aspen trees in the American West, one can find carvings that they left behind as exemplified by those in Figures 29-33. Herders' carvings ranged widely in their intent. Sometimes they were message boards, direction signs, warning signs, boundary markers, or a simple tablet for a poem. Often they were nothing more than simple "doodling" completed in order to pass the time and perhaps memorialize a sentimental thought such as a testament to a past love, a family member, a memorable event, a favorite horse or dog, or even religious belief. Often they were simply markers such as the so well-known "Kilroy was here" meant only to memorialize the fact that an individual had spent a moment of his lifetime in a specific place, particularly ones far from home.

Herders are men who have to be healthy to do their work and were commonly in the prime of life. To believe that they could just leave all their masculine sensual yearnings at home when they went to the sheep ranges is a very foolish and naïve notion. Unless they were gay, they longed for female companionship. The carvings they left behind are, therefore, quite often highly sexual and at times even pornographic in nature, sometimes with considerable ribald but often truly funny humor expressed.

When one comes upon these pornographic images, they should bear in mind that the men who left them most likely never envisioned that anyone else, except for other herders and "woods people," would ever see them. They were seldom meant for the general public

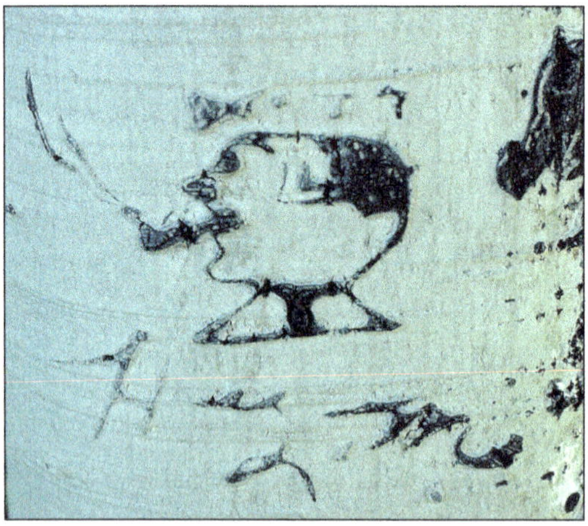

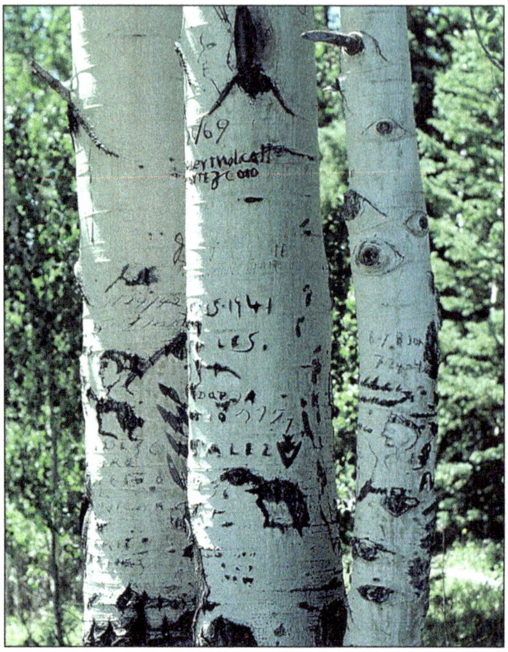

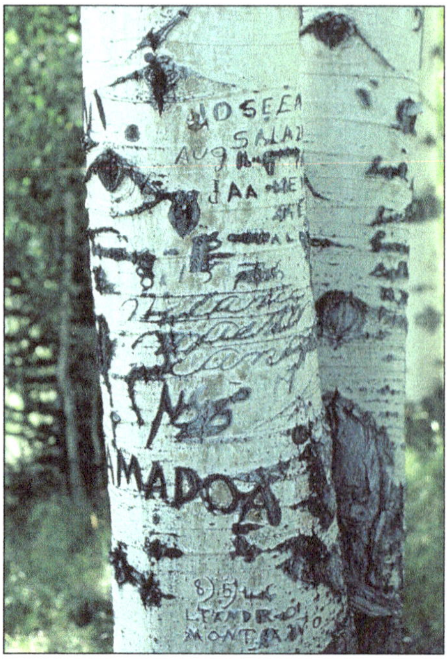

Figure 32: Examples of typical sheepherders' aspen carvings from the Lone Cone vicinity of the Uncompahgre Plateau near Norwood in west-central Colorado (views of 1976 from author's collection).

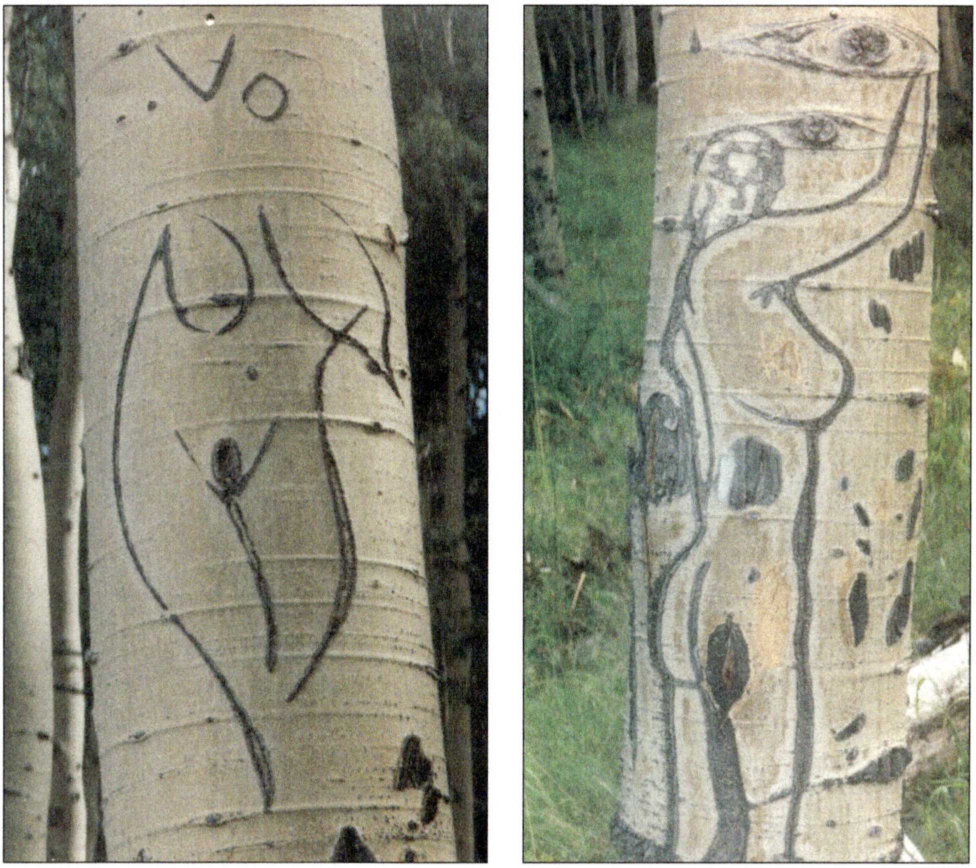

Figure 33: Examples of female figures carved by a sheepherder(s) on Fitzpatrick Mesa near Montrose in west-central Colorado (views of 1995 from author's collection).

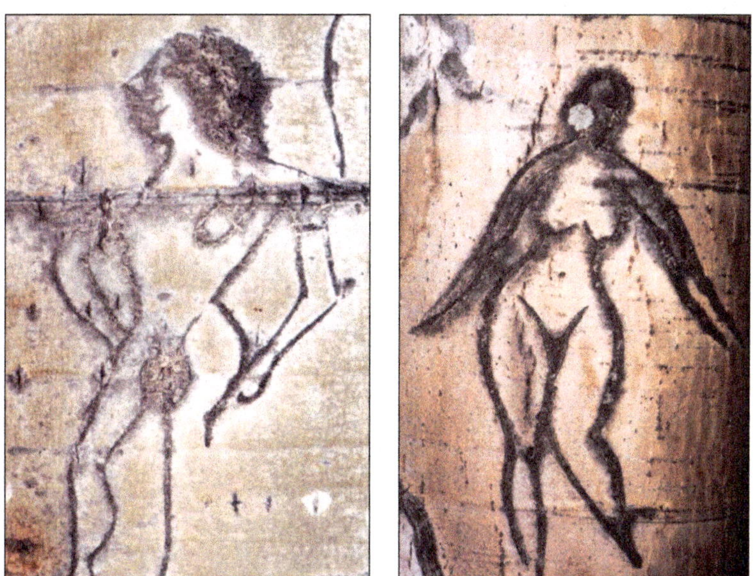

Figure 34: Examples of aspen carvings of women that are considered to be somewhat better than those left by the average sheepherders but which are nowhere near the quality of those left by either Pacomio or Presciliano Chacon (images from the Pagosa Springs vicinity of southern Colorado courtesy of Peggy Bergon, Pagosa Springs, Colorado).

and were a form of sheepherder communication. As illustrated herein, most, if not nearly all, sheepherder carvings are primitive with poor artistic accomplishment, particularly when compared to Paco's work. The images illustrated in this chapter provide some examples of the common form of carvings and lend some comparative perspective on the quality of Paco Chacon's carvings reproduced in his portfolios that follow in Chapters 7, 8, and 9.

I have seen aspen carvings that seem to reflect the mental state of the so-called "crazy sheepherders." Some of these images are so strange that they suggest that the carver had lost all realistic notions about female anatomy and form. With a bit of practice you can begin to read a lot of "armchair psychology" into the herders' aspen carvings. The references and few examples included in this chapter can be consulted for a broader appreciation for sheepherders' aspen art. If the quality of a man's art testifies to the condition of his mental health, then Paco seems to have been psychologically quite healthy!

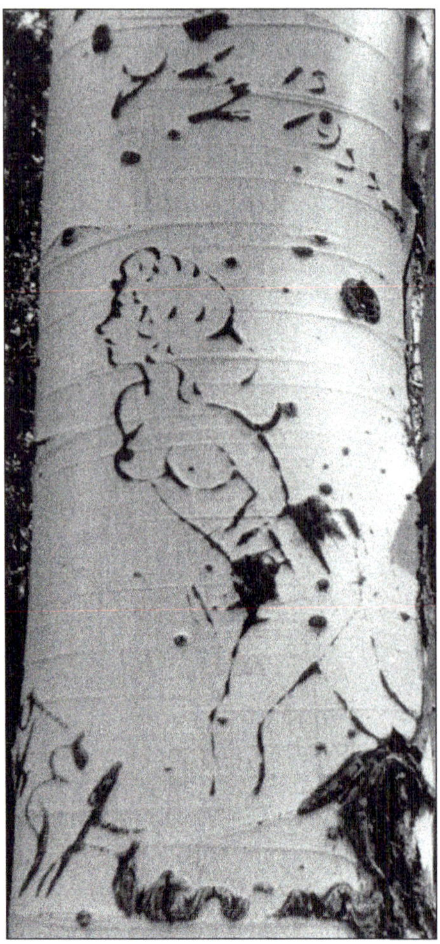

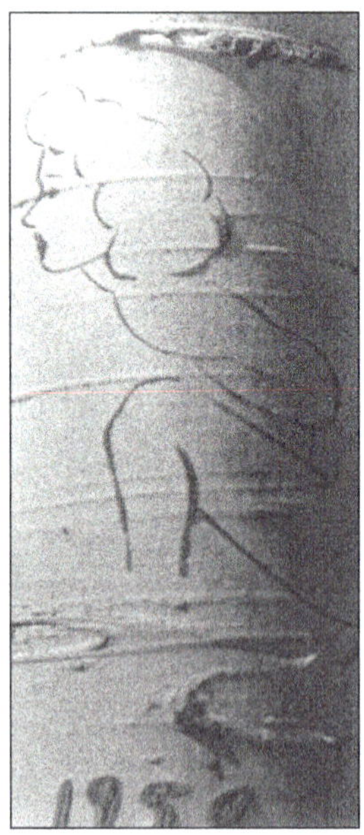

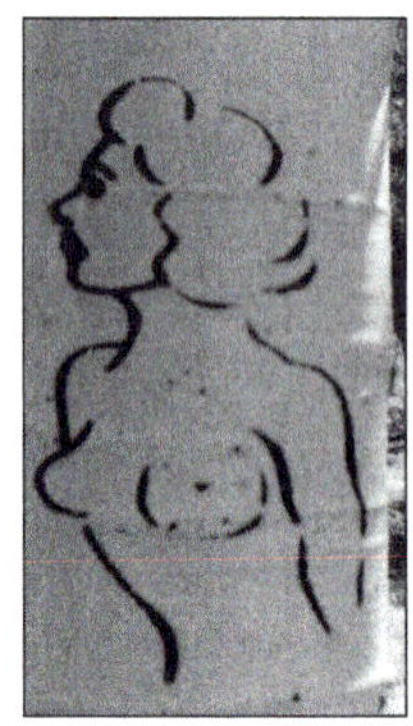

Figure 35: Unsigned examples of sheepherders' well-done aspen carvings from the vicinity of Colorado Highway 90 on the Uncompahgre Plateau in west-central Colorado. These images are believed, primarily on the distinctive and consistent basis of the simple curvilinear contours outlining the hair and facial styling, and particularly that of the nose, to have been carved by Press Chacon, Paco Chacon's younger brother or another member of the Chacon lineage of herders. Images such as these are commonly but incorrectly attributed to Paco. I am unaware of any images signed by him that are at all similar to these. They are, however, very similar to renderings known to have been created and signed by Press. These images are actually second-to-none except for those of Paco himself and attest to the emergent talent that Press also possessed (1960s views courtesy of Monte and Shirley Sanburg, Montrose, Colorado).

Paco Chacon is but one old herder among the many hundreds of thousands who worked in the American West and left their monikers and carvings behind. In this regard Paco's tree carvings are certainly not at all unique. It is the comparatively high quality and prolific nature of his work that sets Paco far apart from these other old "tree writers." A major goal of this book is intended to demonstrate this fact.

It is necessary to acknowledged that Paco's younger brother, Press, also had considerable artistic talent. Press never developed his art to the degree Paco did and was not as prolific as his older brother. But, in fairness to Press, a few examples of his works have been included

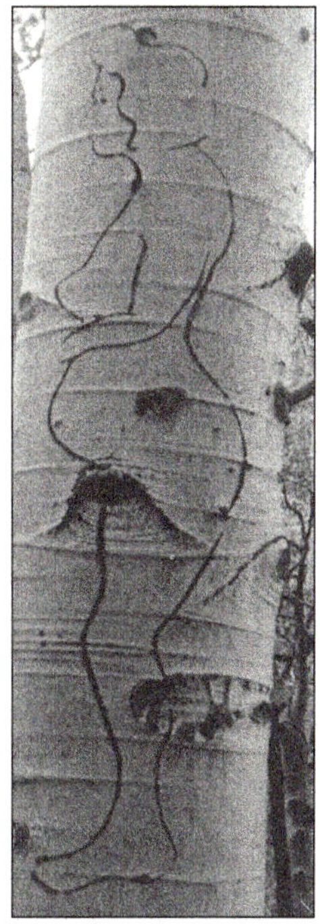

Figure 36: An unsigned aspen carving of a nude and obviously prenant woman from the vicinity of Colorado Highway 90 on the Uncompahgre Plateau in west-central Colorado. This image is very similar to those in Figure 35 and is believed to have been carved by Press Chacon or another member of the Chacon lineage of herders (1970s view courtesy of Monte and Shirley Sanburg, Montrose, Colorado).

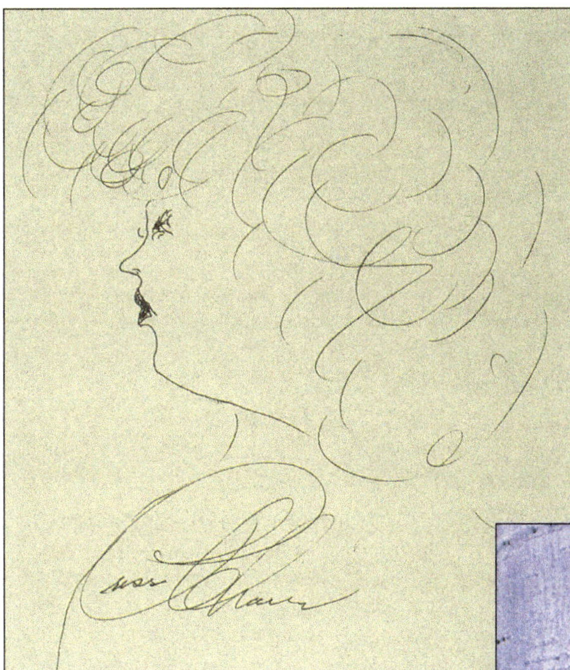

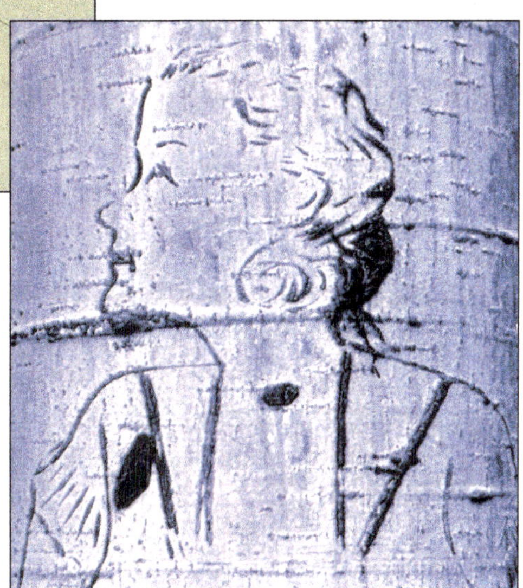

Figure 37: Two portraits of women. The image on the left is a signed example of a drawing by Press Chacon. The right hand image is an unsigned aspen carving from near Lone Cone on the Uncompahgre Plateau of west-central Colorado. While this image might well have been left by Paco, it shows similarities to works by his brother Press. These include the manner in which the eyes and nose were portrayed (author's collection).

(Figures 34, 35, 36, 37). They suggest that the artistic ability best demonstrated by Paco may have run in the family. People sometimes confuse the works of Paco and Press, but with a bit of practice one can learn to differentiate them by the way the women's hair and noses are styled. Press's figures appear to all have been profiles only.

In recent decades the old carvings in the aspen groves of the American West have come to the attention of public land managing agencies and the broader public. They have been fawned over and publicized in books and newspaper articles. Agencies such as the Bureau of Land Management and the Forest Service in many places have active programs intended to record these carvings before the trees die or are harvested for lumber. A simple search of the internet produces a substantial number of entries related to aspen tree carvings.

The State Historic Fund managed by History Colorado (aka Colorado Historical Society) recently provided very substantial grant funds to the San Juan Mountain Association. This grant supported the recording of aspen carvings along the large and historically prominent Pine-Piedra Stock Driveway in southern Colorado. The Basque Studies Program at the University of

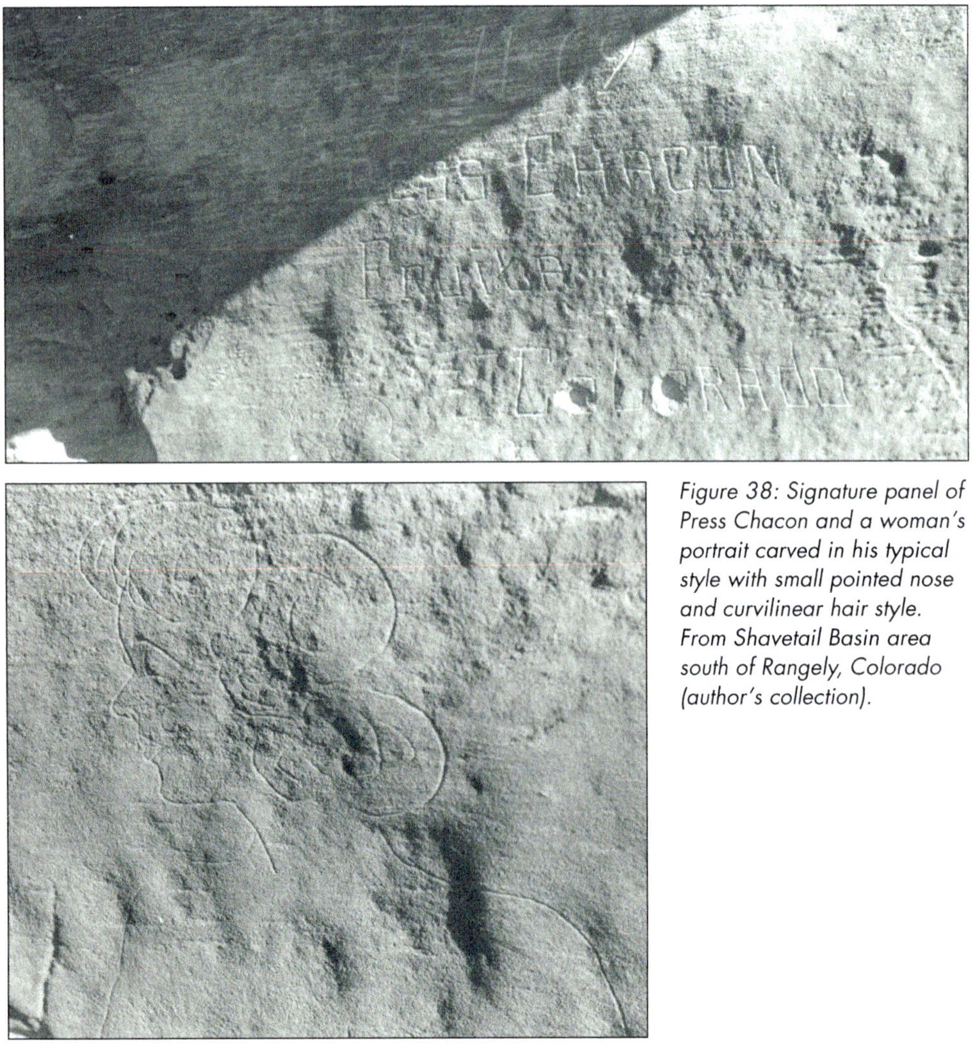

Figure 38: Signature panel of Press Chacon and a woman's portrait carved in his typical style with small pointed nose and curvilinear hair style. From Shavetail Basin area south of Rangely, Colorado (author's collection).

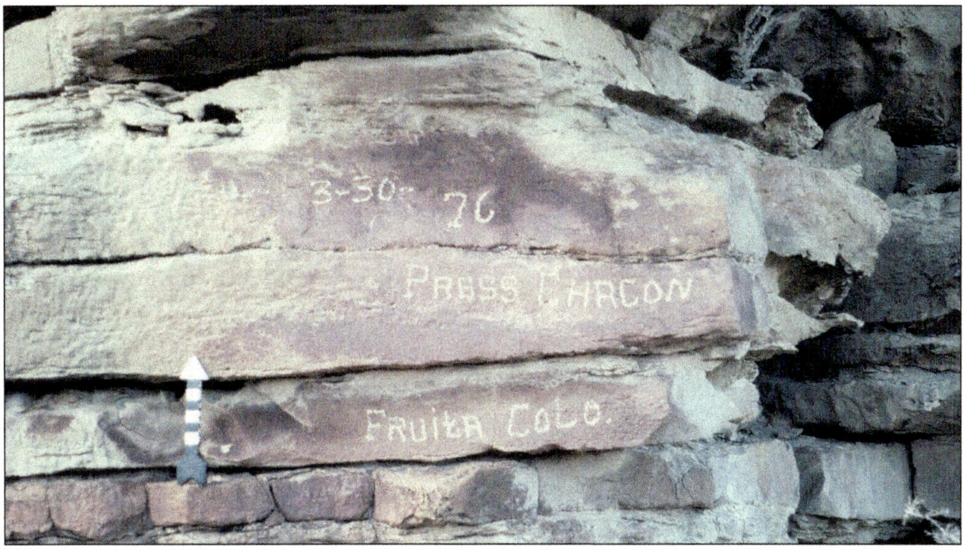

Figure 39: Press Chacon signature panel from Shavetail Basin area south of Rangely, Colorado (author's collection).

Nevada has done much to popularize the aspen carvings left by the Basque herders. J. Mallea-Olaetxe, a professor at the University of Nevada, largely launched his academic career around aspen art and produced one of the first major studies of aspen carving in his well-done volume, *Speaking Through the Aspen: Basque Tree Carvings in California and Nevada* (Mallea-Olaetxe 2000). Although this is a really good book, those few individuals who are familiar with Paco Chacon's work tend to chuckle a bit when the Basque professor extols the beauty of some of the women carved by the Basque herders! The professor obviously never saw Paco's women!

Other published works on sheepherders' aspen carvings include: DeKoorne (1970); Gulliford (2007); Lane and Douglass (1985:68-69); and Sagstetter and Sagstetter (2007). Malakoff (2013) is a particularly good article written from an archaeological perspective. The San Juan Mountain Association recently published its first volume on the aspen carvings of southern Colorado (Lambert 2014). Virtually every land managing agency with aspen forests now has some kind of program in place to record and memorialize the aspen carving sheepherders of old.

Articles on the subject are also commonly appearing in regional newspapers and agency newsletters. Examples include Associated Press (2000); Baker (2004); Colorado Department of Natural Resources (2012); Fishell (1984); and McGonagle (1990). There are also many internet blogs related to the subject such as those by Allen (2008); History Colorado (2014); and Pederson (1997). There are also exhibits such as *In the Great Shadows: Tracing the Sheepherders Legacy* by Earle Swope and Amy Nack (Swope and Nack 2008) and the photographic exhibit of Paco Chacon's work (Chacon and Baker 1992b) first exhibited at the Meeker Sheep Dog Championship Trials in 1992 and then at the Montrose Pavilion the same year. The latter was developed with sponsorship of the Colorado Endowment for the Arts and is now in my possession. I hope it will again be publicly shown in association with the publication of this book, quite likely at the Meeker dog trials or another of the many potential venues.

Although aspen carvings are commonly known and associated with sheepherders, their carvings in stone are much, much less common. The American West contains literally millions upon millions of acres of summer sheep range up in the aspens. Mother Nature thus offered the herders a lot of canvas in these areas. Rocky cliff faces soft enough and suitable for carving are typically not present in the summer ranges.

If cliffs of suitable rock are present, they are usually in the canyons of the winter ranges. Although I have occasionally found what appear to be herder's inscriptions in the rocky canyons of the region, they are quite rare compared to the aspen carvings. This is in large part due to the time consuming labor that is required to complete an image in stone. It is much easier to quickly cut an image into the bark of an aspen than to do so on a rocky cliff face or to carve an entire castle as shown in Figure 28. Although it still takes time to carve an aspen well, anyone can quickly cut into an aspen and leave a poorly executed version of his name or a crude image.

Cliff faces offered Paco good pallets for his work in his winter ranges south of both Rangely, Colorado, and eastern Utah. Paco's rock carvings are formally referred to as "petroglyphs." These are images carved into rock and in this region have very commonly been left by Native Americans who also painted on the rocks. Such paintings are referred to as "pictographs." Paco never painted his images on rocks. All his works are petroglyphs.

In contrast to aspen carvings, the literature of sheepherding is nearly devoid of mentions of rock carving. One can, however, be certain that there are places in the American West, other than Paco's galleries, that bear the monikers and artwork of sheepherders. Famous historic trails such as the Oregon and Santa Fe Trails have distinctive places where people gathered and left their marks on cliffs. In a similar fashion, there have to be other old sheep ranges with herders' rock carvings. But I am unaware of just where they may be.

Those fortunate few who have encountered Paco's works out in western Colorado and eastern Utah have also encountered many other inscriptions and images left by sheepherders, particularly among the aspens. Their great appreciation for his work is therefore not, as it often times is with some individuals, a result of overly exuberant naïveté, ignorance, or a lack of information about sheepherder art. Like me, they have spent their careers dealing with it among other forms of archaeological remains. They have a very good frame of reference within which to refer to Pacomio Chacon as the "master sheepherder artist."

Others who have written about sheepherders' aspen art have not treated any of Paco's works, and it appears that they have never seen them. They are far from population centers in areas that are still today relatively distant, desolate, and not commonly frequented by hikers and tourists. I am certain that if they had seen his works, the admiration for most of the other herders' art would be somewhat less effusive than it presently is.

Paco Chacon left his art and his signature on trees and cliffs from near the Continental Divide in Colorado west into eastern Utah and from the north slope of Colorado's San Juan Mountains north into northwestern Colorado and northeastern Utah. It may even extend into southern Wyoming and northern New Mexico.

Paco was obsessive about his carving. Although we have no idea of how many carvings he left behind, it must have been in the many hundreds if not thousands. Anywhere he worked, he left his art if there was anything at all to carve on. Although he always stressed that he never thought much about his art, Paco knew his work was good in terms of what it was, namely folk art. Paco was a folk artist. If there was ever another carver as good at working in these mediums as Paco, my colleagues and I are completely unaware of it, despite many years of concerted research on the subject.

By nearly universal definition, the term "folk art" refers to artistic work typically produced in cultural isolation by untrained, often anonymous, artists or by artisans of varying degrees of skill. Folk art can include pottery, basketry, painting, carving, and a rich variety of other items and mediums. I am certainly no art critic and cannot even recall taking a university class in art appreciation. At one point in the research for this volume, I contacted some of the major art auction houses, including Christies, to find out whether there was any truth to the rumors they had sold some of Paco's artwork.

Although there was no resolution to the matter, after seeing Paco's work, the specialists I consulted all agreed that Paco's work was quite good and belonged to the genre that in recent decades has become known as "outsider art." By modern critical definition, outsider art is the work produced outside the mainstream of modern western art by:

> *self-taught, untrained visionaries,*
> *spiritualists, eccentric recluses,*
> *folk artists, psychiatric patients,*
> *criminals and others beyond the imposed*
> *margins of society and the art market*
> (Rhodes 2000; also see O'Connor 2005).

Over the years Paco's art has been recognized in a few local newspaper articles and through interviews conducted by the Forest Service. As his biographer and spokesman, I have spoken about Paco's work before archaeologically oriented gatherings (Baker 2003a). In 1992 Paco received significant recognition when the Colorado Endowment for the Arts awarded funds to the Museum of Western Colorado to sponsor our travel to the Meeker Sheep Dog Championship Trails in Meeker, Colorado, so that he could demonstrate and speak about his talents (Figure 40).

That grant application noted that Paco was considered by the museum to be a "master folk artist" (Perry and Sharpe 1992). The grant also paid for creation of the formal exhibit of his work, which was shown at the dog trails. At that time the endowment agreed that Paco was a "master" folk artist. Paco and I subsequently appeared on Colorado Public Radio (Rumsey 1992) to talk about his work, and other people soon interviewed him.

In 2003 the Museum of Western Colorado teamed with my firm, Centuries Research, Inc., and applied for a grant from the State Historical Fund administered by the Colorado Historical Society. That application was supported by the Rocky Mountain Regional Office of the U.S. Forest Service and some of its local forest offices, which administered lands where Paco left carvings. The grant request was for support to "better understand, record, manage, and preserve" Paco's unique aspen carvings. In the application the museum and forest service recognized Paco as "a master folk artist in the medium of arborglyphs." The agencies held a view that through his aspen carvings Paco's contributions to the cultural landscape of western Colorado were "significant" (Hunt 2003; Ketelle 2003; Overturf 2003; Perry 2003).

Unfortunately, the State Historical Fund did not believe it could support the project unless it was part of a larger, more comprehensive effort to study aspen carvings. The agency stated that it could not justify funding a study of the work of only one sheepherder artist. I suspect that his penchant for carving nude women fed into this negative decision by a state agency. Although I had kept the project alive and had been slowly chipping away at it, no further substantial efforts were made to further study Paco's work until 2013 when I finally resolved to complete this book in the near future. At that time I began working more closely with Paco's family and commenced one last-ditch effort to gather as much additional material as I could.

Now that my readers have been introduced to Pacomio M. Chacon and his sheepherding lifestyle and artwork, it is my great honor to here finally offer them an opportunity to view his art. Many of Paco's secrets from the remote and far-flung canyons and forests have finally been gathered herein! No longer will they be appreciated by only a select few who roam in these kinds of places! Let us first tour Paco's aspen art (Figure 41).

Chapter 7

PACOMIO'S ASPEN ART

Aspen carvings are a very impermanent art form that can survive only as long as the tree on which they were carved still stands with its bark intact. Once the tree dies and falls, the images decay along with the bark. Moreover, some more subtle issues are involved in the impermanent nature of some aspen carvings than just the lives of the trees. Well-executed images pass through a natural life cycle of their own. This consists of at least five stages. For well-done, expertly carved images, there is only a brief period when they truly bloom and appear at their very best before beginning to deteriorate. They are like annual flowers that only blossom for a short period before they die.

The five stages in a carving's lifecycle only apply to well-executed images such as Paco's. They do not readily apply to the poorly executed images typically left on aspens by sheepherders and other people. These are often literally hacked or roughly gouged all the way through the

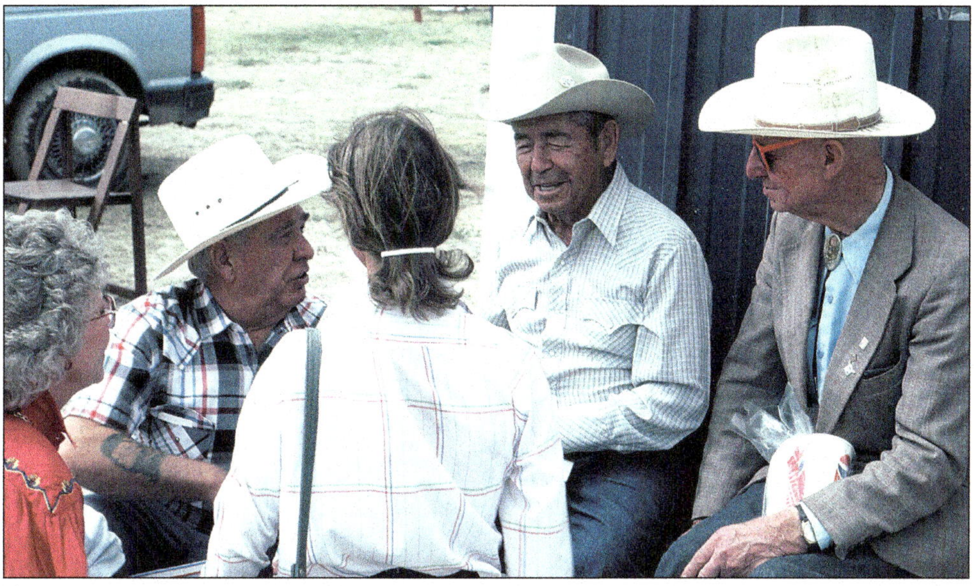

Figure 40: Paco Chacon (center) discussing his aspen carving with admirers at the Meeker Classic Championship Sheepdog Trials at Meeker, Colorado, in September 1992 (author's collection).

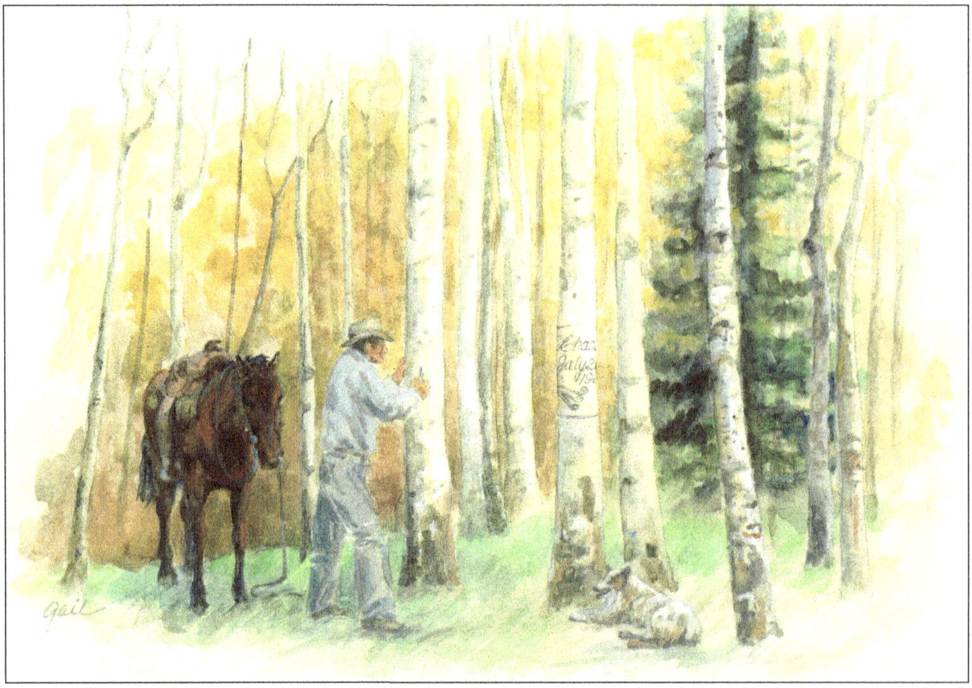

Figure 41: Artist's conjectural image of Paco Chacon carving an image on an aspen tree (watercolor painting is by Gail Carroll Sargent of Centuries Research, Inc., Montrose, Colorado).

bark and deep into the underlying tissue. Such extreme carving never allows an image to enter full bloom before rapidly beginning to scar over. The more severe the cuts, the faster and more vigorously the tree will respond and attempt to heal the wound by producing heavy scars that quickly obliterate the image.

With Paco's images and those of other master carvers, the first or nascent stage in the life of a carving commences with the selection of a healthy aspen tree that is large enough to carry the envisioned image. Ideally, the tree also needs to have smooth, clean, and white bark. The best carvings were usually completed between July and about mid-September. In the high mountain country where the aspens grow, the trees are preparing themselves for winter by September. According to Paco, after that date the trees never produce a respectable image. Paco's powers of observation of nature were keenly developed, and he understood the life cycle and seasonal nuances of aspens. He could actually predict how individual trees would react to being carved. Paco spoke of this seasonality and the trees' responses.

> *They carve the tree and it change like people, you know? In July through part of September it is good to carve but before or after that it no good ...the barks is sort of like people, the tree change and get ready for winter and the bark grow deeper, I mean higher, I mean thicker. The bark is thicker in the fall as to get ready for winter.*

Paco carved at any time of day that he had an opportunity from about July into September. After mid-September the bark becomes too tight because of the cold. Both Paco and Press agreed that their lives as sheepherders gave them "all the time in the world to carve."

Paco typically began a carving by drawing it on the white tree bark with the stub of a lead pencil he always kept in his pocket. He then very delicately followed the penciled image with the finely sharpened tip of the smallest blade of his pocketknife. This procedure separated him from most other carvers. Paco once stated that

> *A lot of people ruin the trees.*
> *They don't know how to cut with a small knife.*
> *I do it with a tiny little knife, little bitty*
> *knife, and it takes little pressure and*
> *don't cut too deep.*

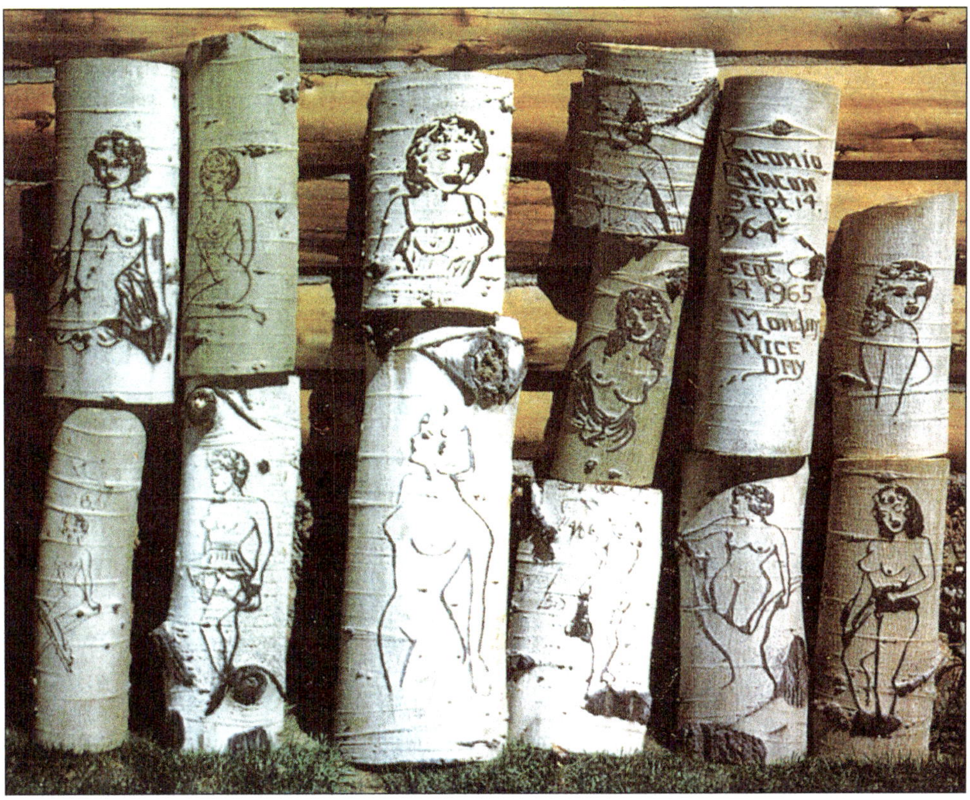

Figure 42: A portion of Paco Chacon's Stage 2 and 3 aspen art collected in the ca. 1970s and 1980s by the logger, the late Richard Moyer (1926-2015), from the White River National Forest near Meeker, Colorado, while they were in their florescence. This view of only a part of his collection was assembled on his front porch and photographed with a Polaroid camera by Moyer years ago. Whatever portion of this collection remained in his possession at the time of Moyer's death was either given away by him–and he was known to have been generous with Paco's carvings–or was stolen from his barn (original photo courtesy of Dick Moyer of Meeker, author's collection).

Paco's tediously executed carvings were completed with very shallow cuts that only penetrated through the outer bark of the tree down just barely into the bright green of the fresh cambium beneath it. Deeper cuts caused heavy scarring and splitting of the bark. Once these fine, shallow cuts were made, they were nearly invisible. One would probably not even notice their presence unless they were already aware that an image had been carved on the tree. There was no chiseling or gouging of the lines to reveal an image immediately, a common practice of most carvers. Carving of the fine lines completed Stage 1 in the life of one of Paco's aspen carvings.

Stage 2 in Paco's carvings commenced after his completion of the fine cuts and lasted for two to three years. During this period the tree began to heal the fine cuts, and they began to fill with black scar tissue that does not quickly overgrow the image. The images became more and more visible and were still evolving. At the end of this period, the fine lines became readily evident but were never static. The trees were still alive, and the process of healing and scaring continuing.

Stage 3 witnessed the full bloom of the images when they appeared at their very best. Paco's understanding of this evolutionary process was what really separated him from most other tree carvers. Most people who carve aspens want the gratification of immediately seeing their completed images. Paco was content to wait. He knew it was just a process of the passage of time. Occasionally he was unable to complete an image to his satisfaction because the sheep moved or something else intervened, and he had to interrupt his carving. This is why some of his carvings are unsigned. Often he never even saw his works in their full glory. His duties as a sheepherder could prevent him from returning to the locations where he had carved. He understood the trees and knew that the images would appear in their own good time. In

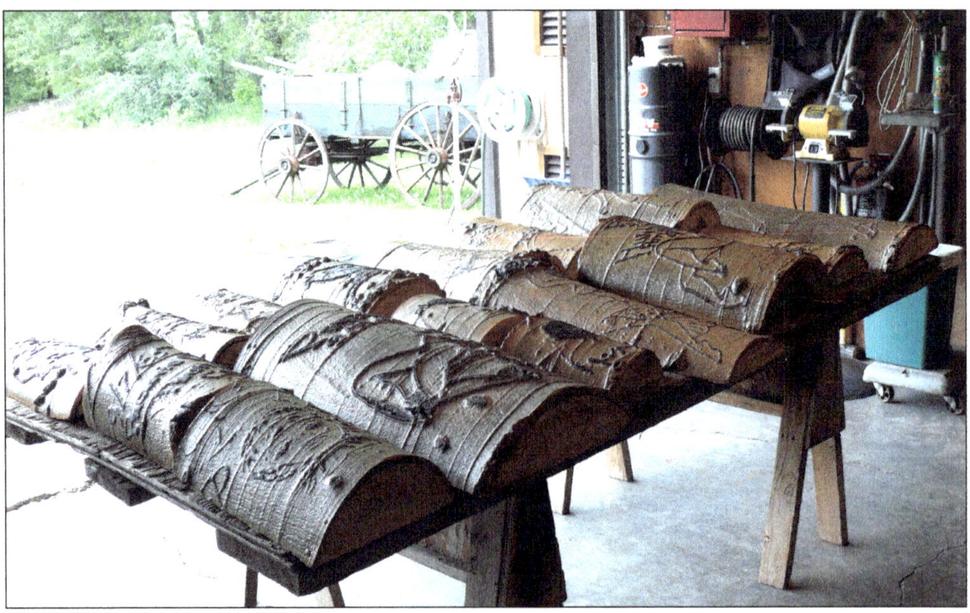

Figure 43: Some of the Cogswell collection of Pacomio Chacon's aspen art collected from the White River National Forest near Meeker, Colorado (photo opportunity courtesy of Dave Cogswell of Meeker).

Stage 3 Paco's images fluoresced for a brief period, perhaps just a few years, before Stage 4, or obliteration, started to set in.

After Paco's original cuts were made, the trees continued to heal them throughout their lives. Scarring of the cuts continued until thick, black crusts began to develop in Stage 4 and started to obscure the lightly scarred fine lines of the original images. Just how long this took to begin to significantly degrade the fully fluoresced images of Stage 3 is not known to me, but it cannot have been all that long. To the best of my understanding, this occurred within a decade or two of their inception. Some of the images in Paco's portfolio herein evidence this deterioration when compared to the fresher Stage 3 images. Stage 5 in the life cycle of an aspen carving occured when the tree died and ultimately crashed to the forest floor, was burned in a forest fire, or was harvested for lumber or to obtain Paco's images.

Paco's subjects consisted primarily of nude women, but he also carved just his name at times, as well as horses, deer, elk, men, cowboys, and men on horses. He could not remember ever carving a religious symbol on a tree. According to Paco there were no special purposes in his selection of subjects. A lot of his carvings, with exception of his women, were based on his surroundings and things he was noticing within them.

In his later years, enough people knew Paco's work that a few came to him and paid him to carve on an aspen. They eventually cut the trees and took the logs with Paco's images into their homes for display. Lamps and bookends have been made from them, and a couple of families in the Meeker area amassed substantial collections of Paco's artwork (Figures 42, 43). People who have obtained samples of Paco's works prize them highly and are usually unwilling to part with them. Paco was always hurt and offended that people were taking his carvings from the forests. He wanted them left there for others to enjoy.

A logger in Meeker, Richard Moyer, amassed a very substantial collection of Paco's aspen art that was present in government timber sales. This individual was also an outfitter who catered to wealthy hunters from out of state. It was not unusual for him to give a sample of Paco's work to favored clients. By this means examples of Paco's work have spread far and wide but cannot be traced. Figure 42 illustrates a portion of this individual's collection of Paco's work. Figure 43 shows another local collection of Paco's work from the Meeker area. It is very fortunate that these individuals collected these pieces. Without these collections there would be very little record of Paco Chacon's aspen art still surviving. I have personally seen very few of Paco's aspen carvings in the forests. Most of those I have seen have come from existing collections or photos passed on to me.

PACO'S ASPEN PORTFOLIO (PAP)

The following portfolio of Paco's aspen images contains photographs taken by several people over many years with a variety of cameras, film, and digitization. Some logs bearing carvings have also been treated with various preservatives that has made them shiny. The images are, therefore, not at all consistent in their color or quality. I would remind the reader that this was very much a "salvage project" intended to gather and record examples of as much of Paco's work as possible before it was gone completely. I cannot even begin to guess what percentage of his carvings have been recorded herein. It is probable that it is only a minor part of his "woodland portfolio." Many were probably lost decades ago.

The images offered here have been grouped into two sections. The first or Group One, contains his highest quality images, which were typically collected and/or photographed at or near their fluorescence in Stages 2 or 3 of their lifecycle. These images are followed by Paco's Group Two images, which are of lesser artistic quality; he completed them very quickly; or they were in Stage 4 and well past their fluorescence when collected and/or photographed. This group of images contains carvings that were badly scarred over by the time they could be photographed. Unless otherwise noted, all images are from the White River National Forest near Meeker, Colorado, where Paco is known to have herded on the summer ranges for Mike Theos in the 1970s and others prior to that period. The dates when carvings were photographed or collected were seldom, if ever, recorded.

PACO'S GROUP ONE ASPEN CARVINGS

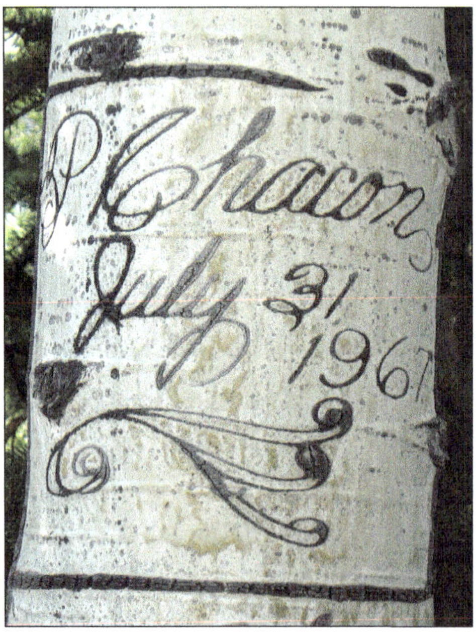

PAP 1 (Paco's Aspen Portfolio 1): Classic elaborate signature from the White River National Forest.

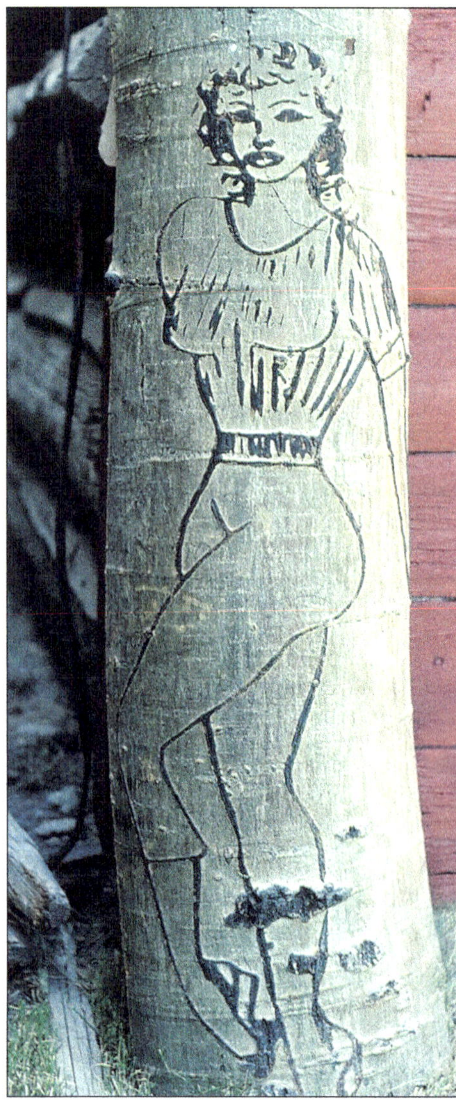

PAP 2: Paco's "Aspen Lady" (photo opportunity courtesy of the Meeker Historical Society Museum, Meeker, Colorado).

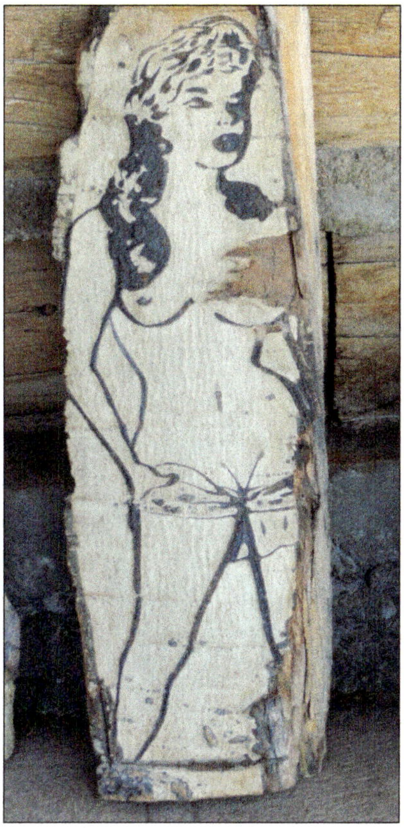

PAP 3: Paco's "Hi Fellas!" (photo opportunity courtesy of Richard Moyer, Meeker, Colorado).

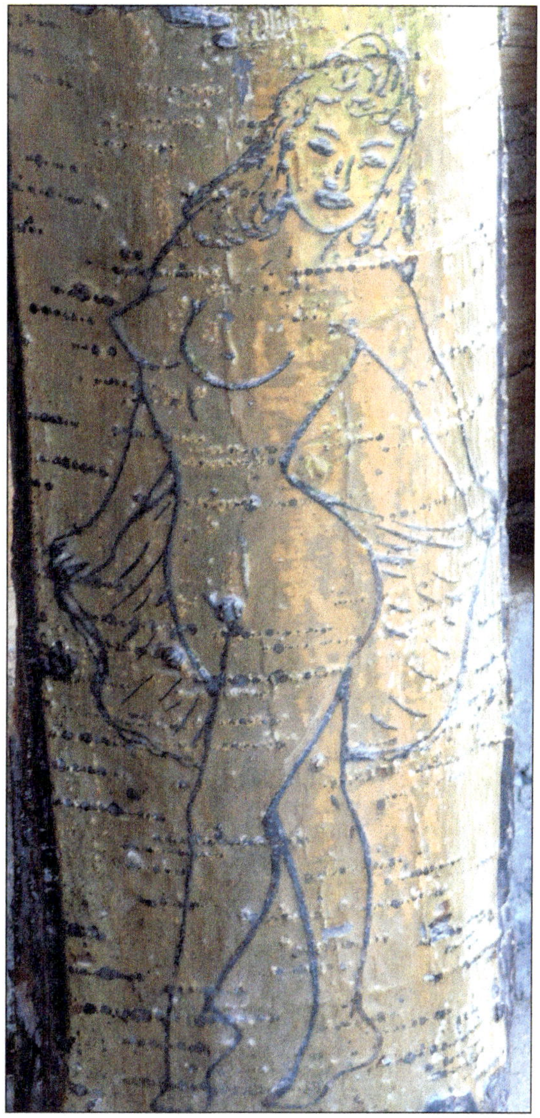

PAP 4: Paco's "Flouncy," one of Paco's very finest aspen carvings (photo opportunity courtesy of Richard Moyer, Meeker, Colorado).

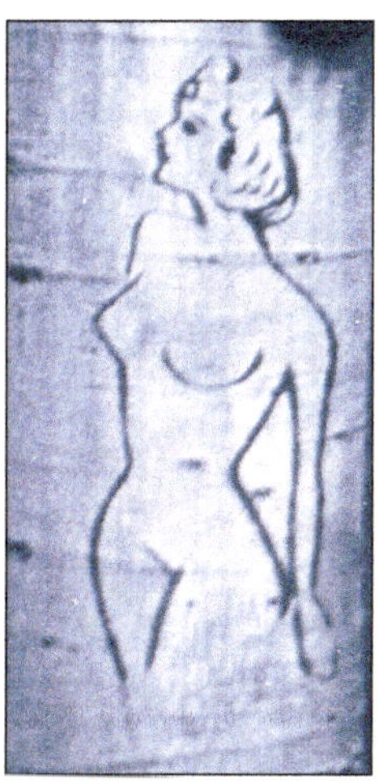

PAP 5: Paco's "Marilyn Monroe the 2nd" (photo opportunity courtesy of Richard Moyer, Meeker, Colorado).

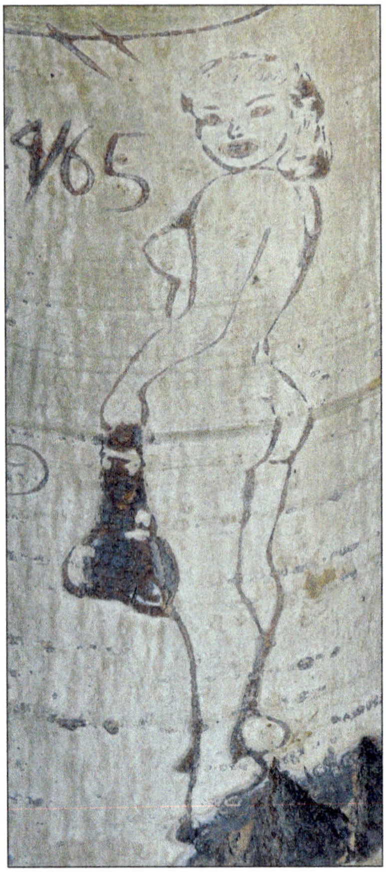

PAP 6: Paco's "Little Traveler" waiting for the bus (photo opportunity courtesy of Richard Moyer, Meeker, Colorado).

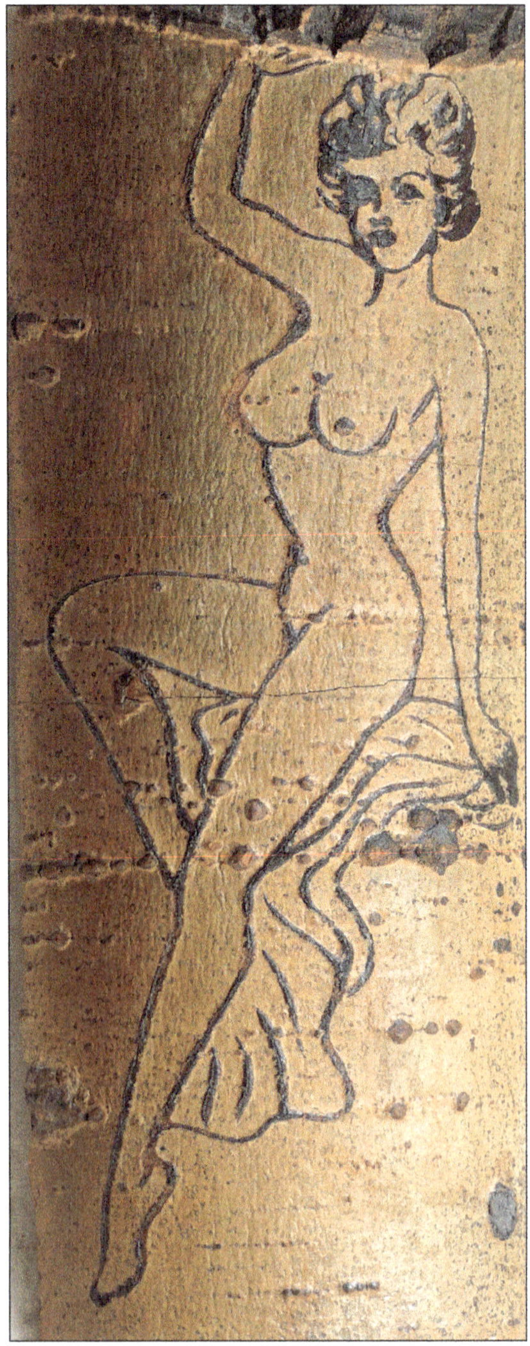

PAP 7: Paco's "Forest Charmer" (photo opportunity courtesy of Dave Cogswell, Meeker, Colorado).

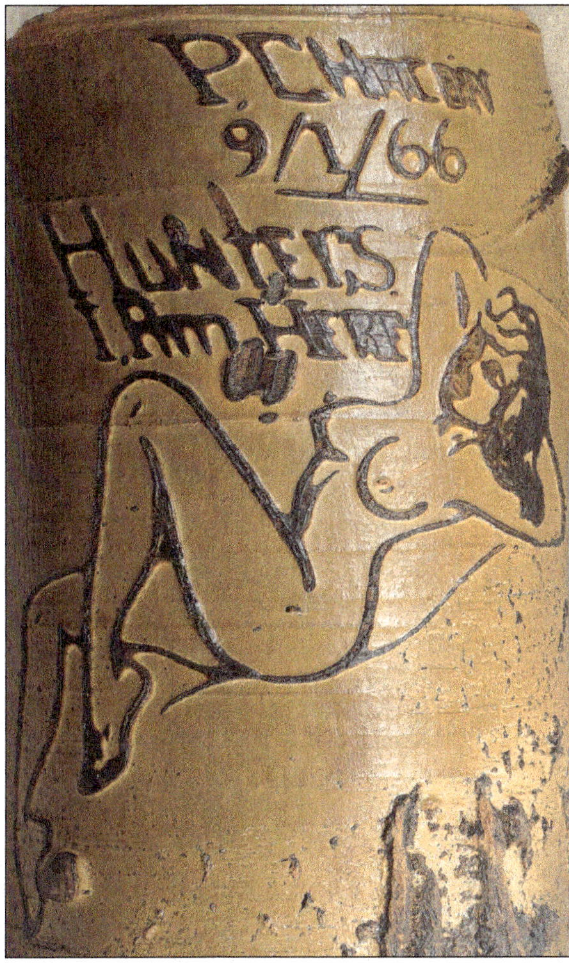

PAP 9: Paco's "Hunters Welcome." The story behind this image is believed to derive from the fact that prostitutes in the years after World War II were known to frequent the hunters' base camps in the Colorado mountains during hunting season. I have heard of these accounts from Paco and other individuals over the years. These women were sometimes known to travel about in old buses (photo opportunity courtesy of Dave Cogswell, Meeker, Colorado).

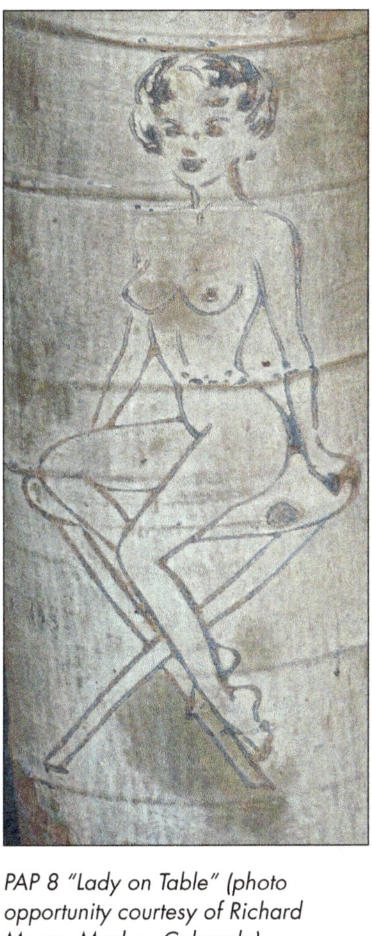

PAP 8 "Lady on Table" (photo opportunity courtesy of Richard Moyer, Meeker, Colorado).

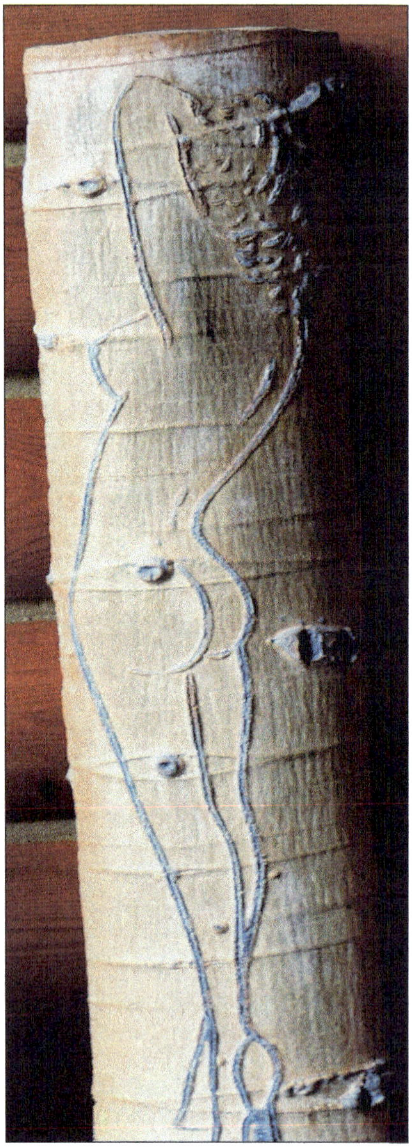

PAP 10: Paco's "Morning Stretch" (photo opportunity courtesy of Richard Moyer, Meeker, Colorado).

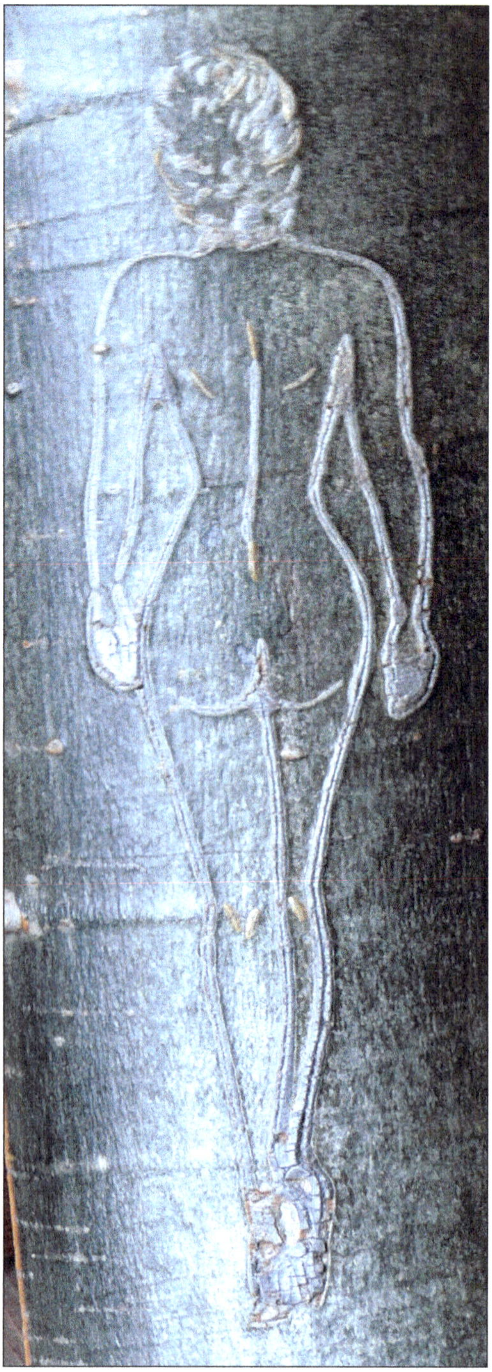

PAP 11: Paco's "Study in Retrospective" (photo opportunity courtesy of Richard Moyer, Meeker, Colorado).

PAP 12: Paco's "Morning Eyes" (photo opportunity courtesy of Richard Moyer, Meeker, Colorado).

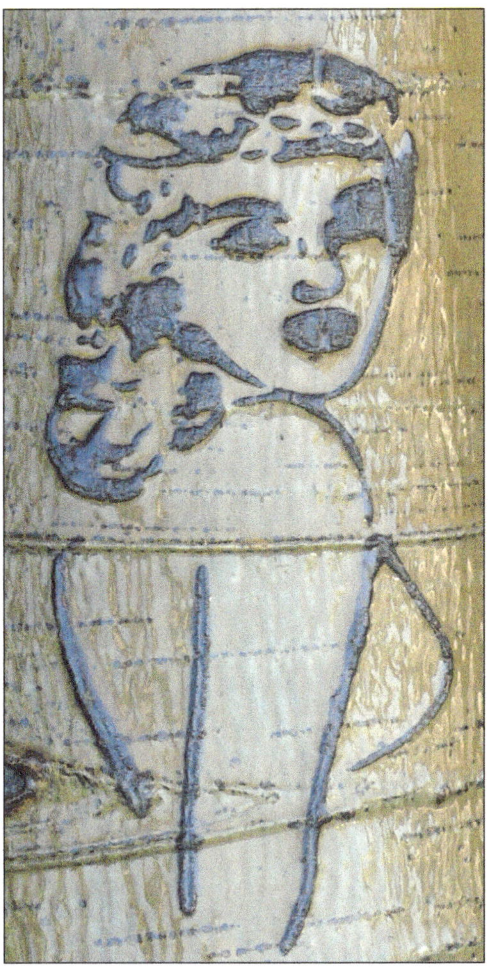

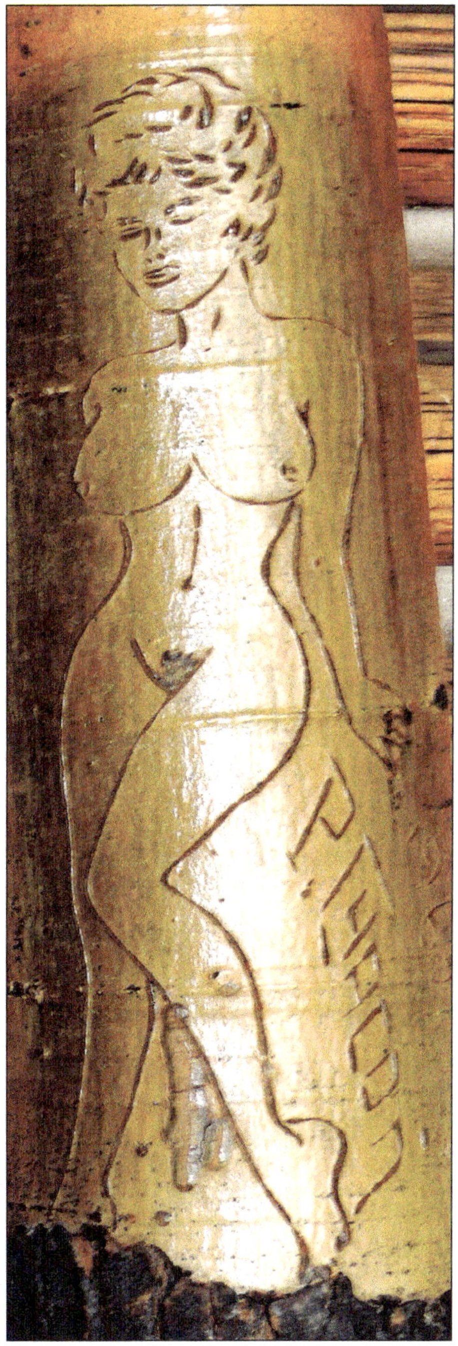

PAP 13: Paco's "Woodland Dream." This is yet another of Paco's finer works (photo opportunity courtesy of Richard Moyer, Meeker, Colorado).

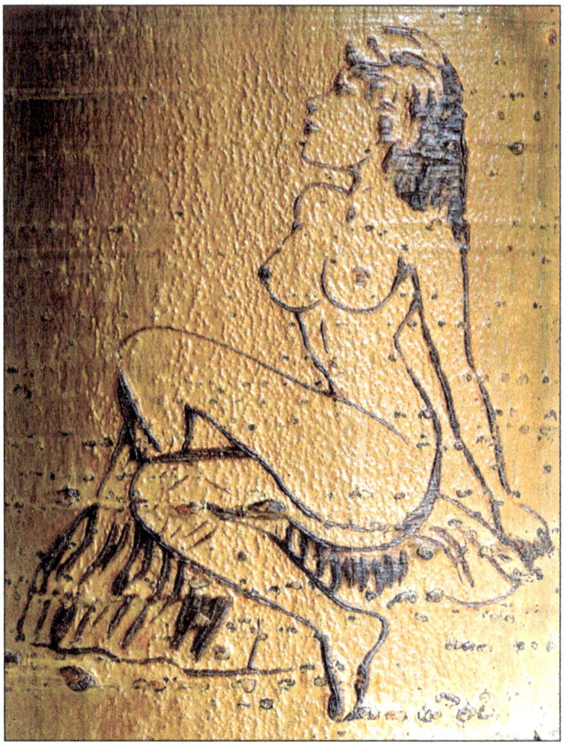

PAP 15: Paco's "Lady a Sunning" (photo opportunity courtesy of Dave Cogswell, Meeker, Colorado).

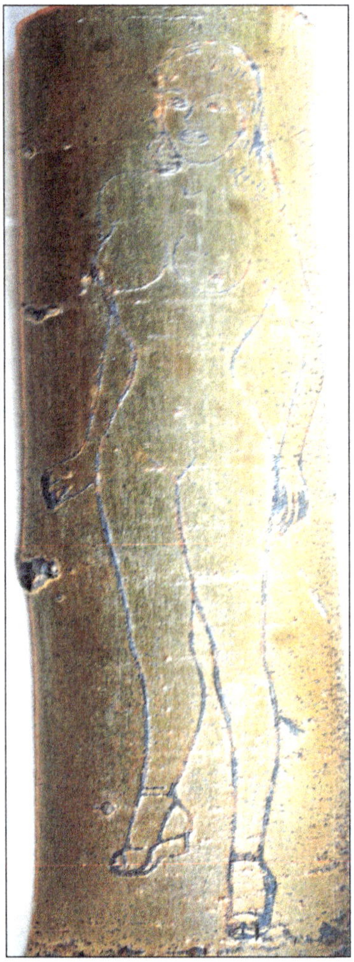

PAP 14: Paco's "Aspen Treat" (photo opportunity courtesy of Richard Moyer, Meeker, Colorado).

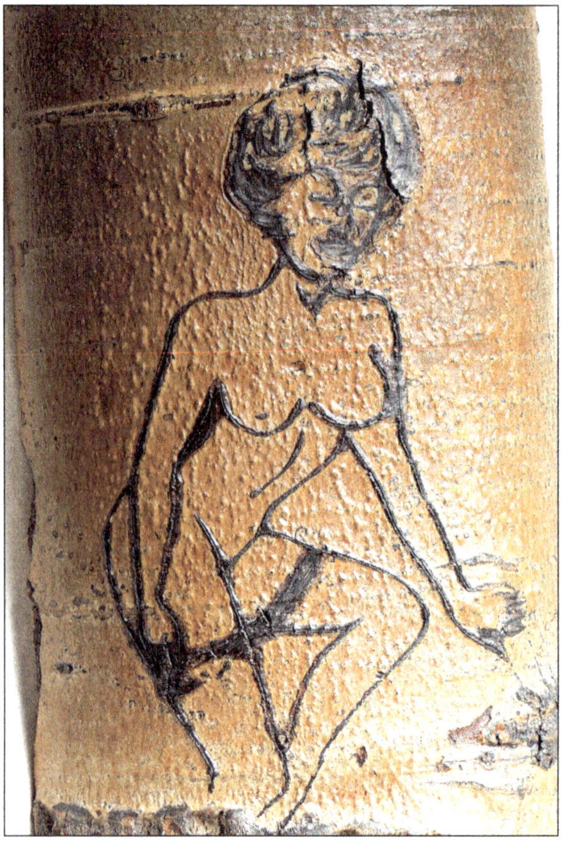

PAP 16: Paco's "Little Stockings" (photo opportunity courtesy of Dave Cogswell, Meeker, Colorado).

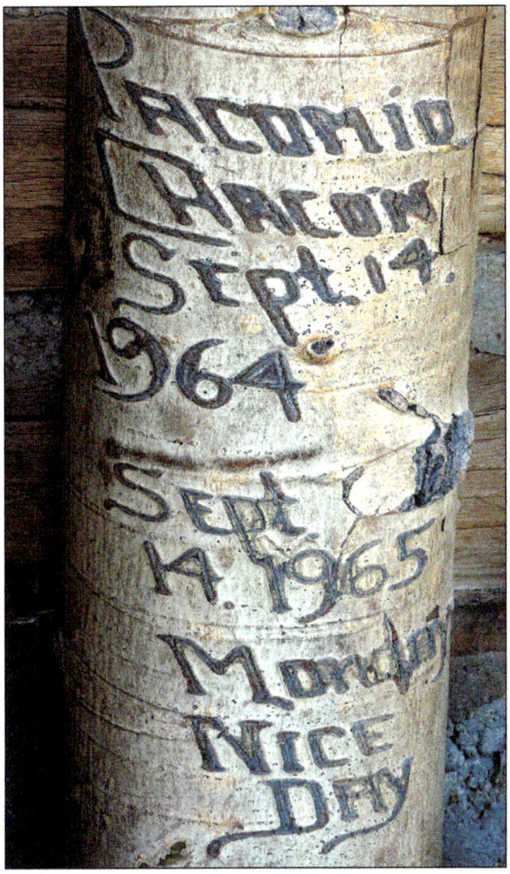

PAP 17: Paco's signature (photo opportunity courtesy of Richard Moyer, Meeker, Colorado).

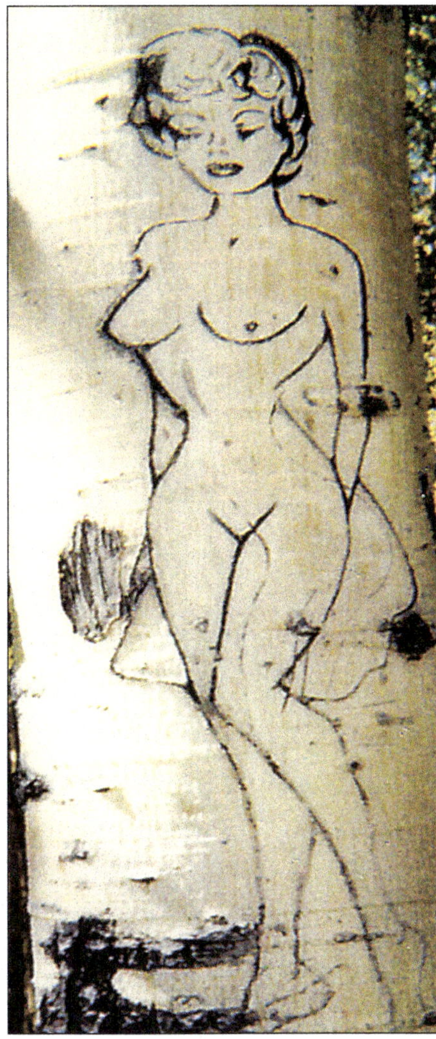

PAP 19: Paco's "Marilyn M.-Chimera of the Aspens." Another of his finest works.

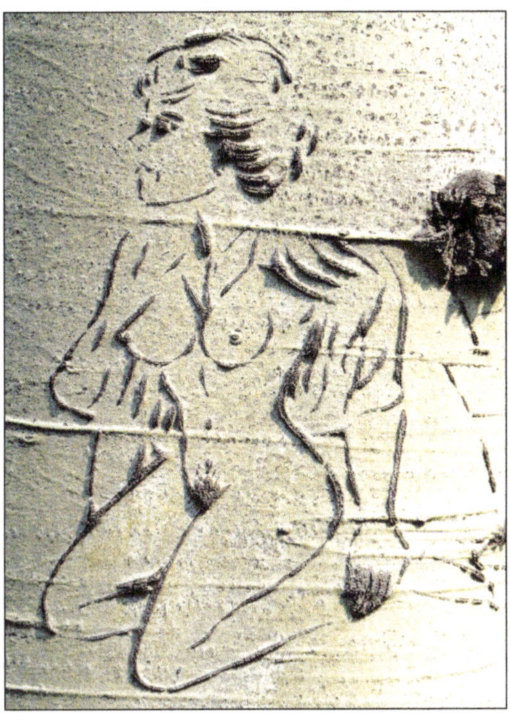

PAP 18: Paco's "Kneeling Lady.

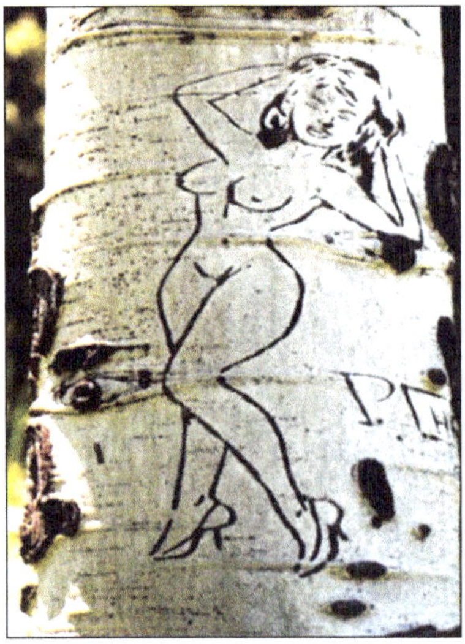

PAP 20: Paco's "Little Disco Dancer."

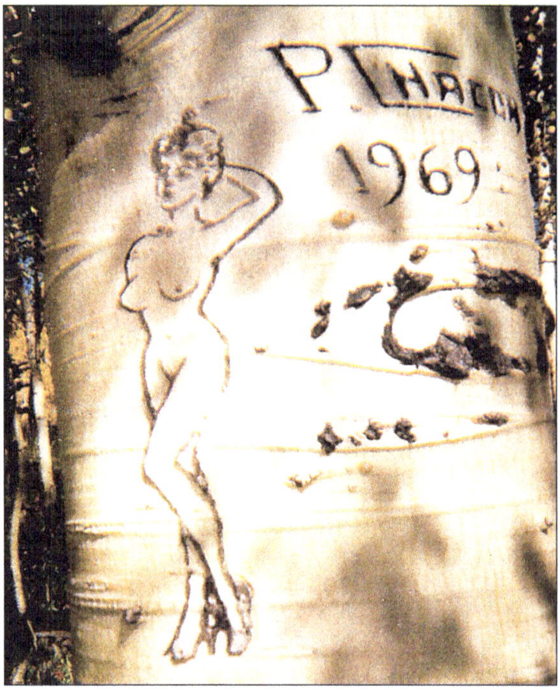

PAP 21: Paco's "Welcoming Lady."

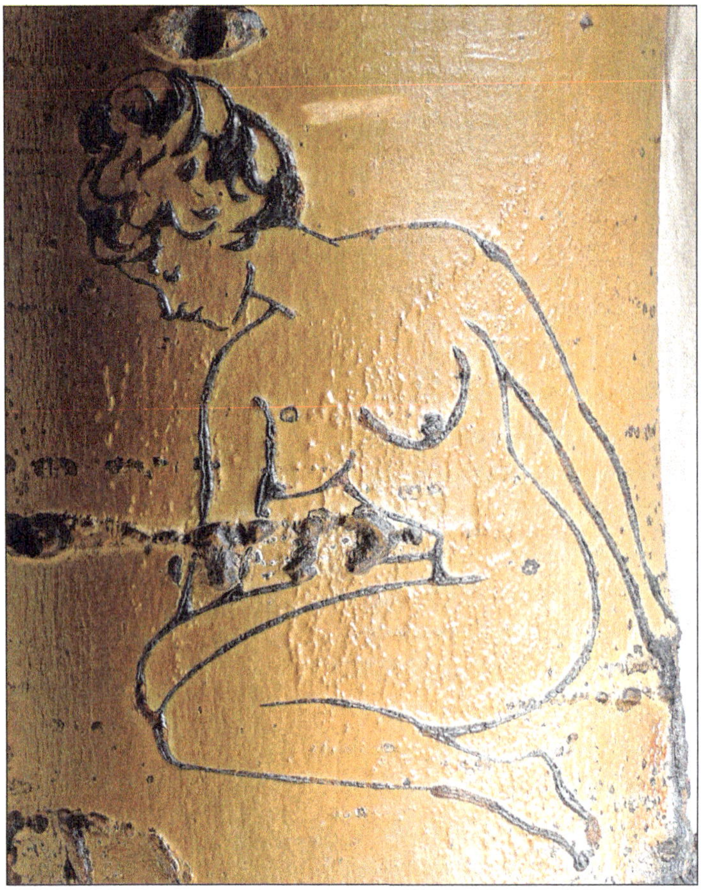

PAP 22: Paco's "Lady a Gazing" (photo opportunity courtesy of Dave Cogswell, Meeker, Colorado).

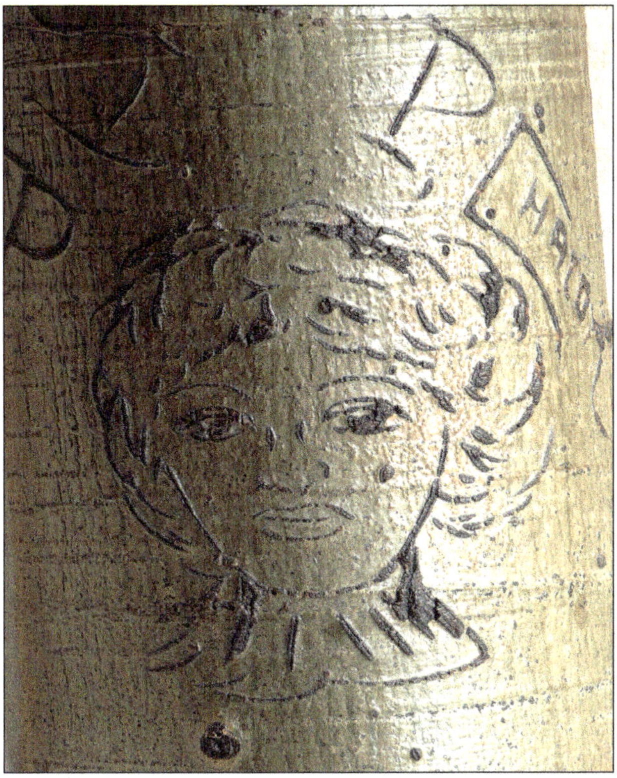

PAP 23: Paco's "Portrait in Aspen" (photo opportunity courtesy of Dave Cogswell, Meeker, Colorado).

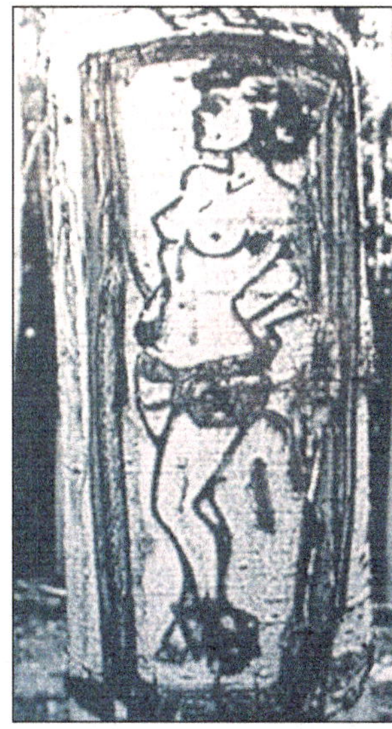

PAP 25: Paco's "Whimsy Girl." This is one of the earliest known female images made by Paco and likely dates to the early 1950s (from outside the kitchen door at Arrowhead Resort, Uncompahgre National Forest, Jim Houston Collection, Gunnison, Colorado).

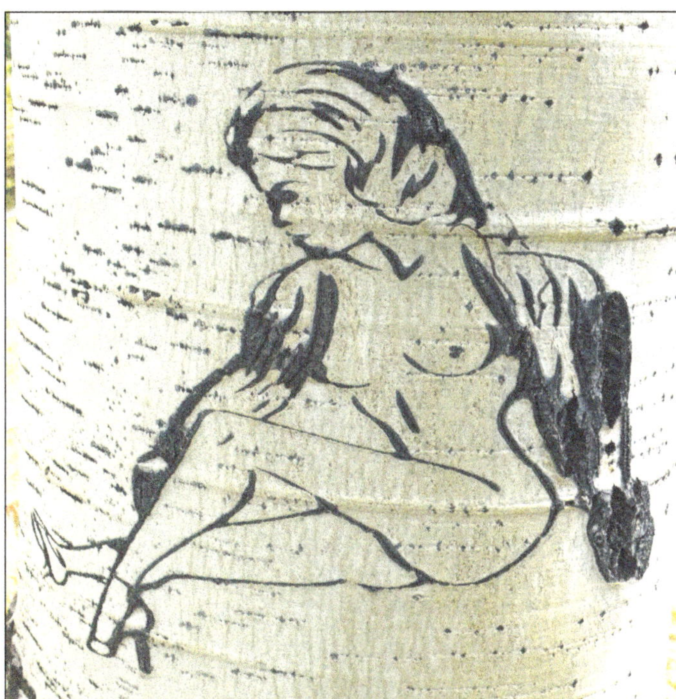

PAP 24: Paco's "Lady with Shawl."

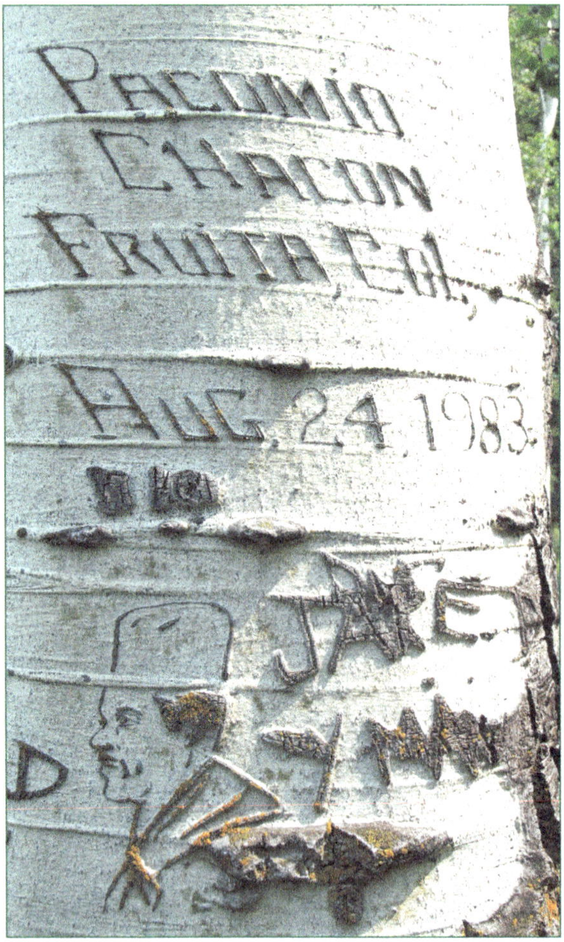

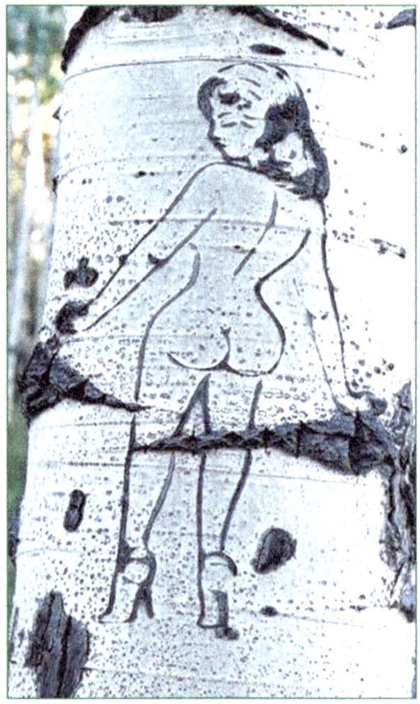

PAP 26: Paco's "Little Bashful Cheeks."

PAP 27: Paco's "Boss Man" from 1983.

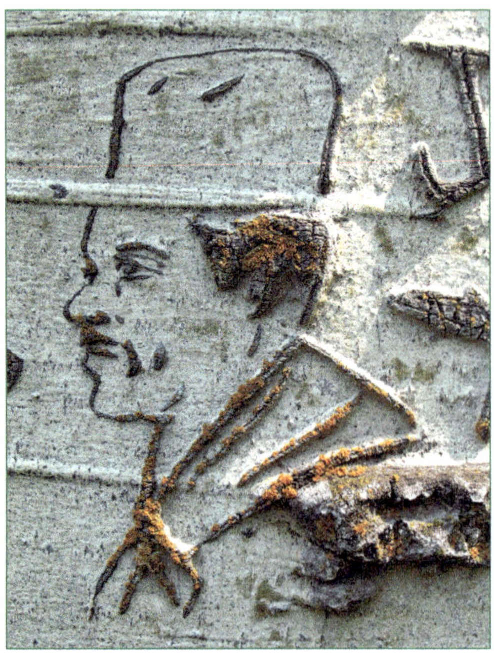

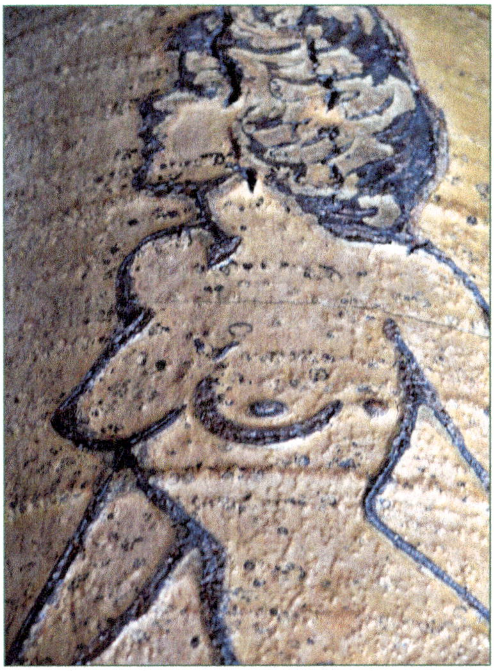

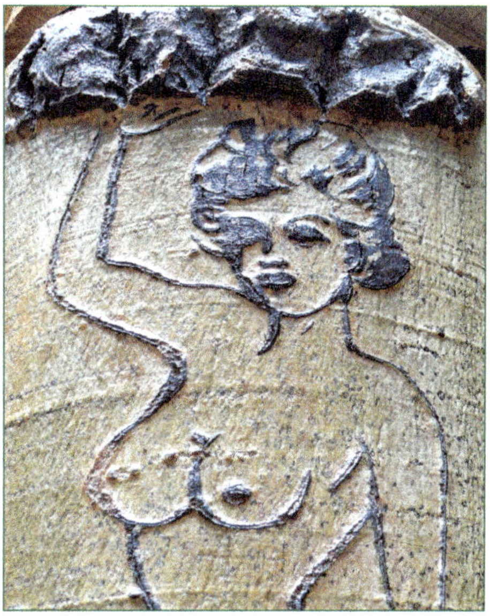

PAP 28, 29, 30: Details from Stage 2/3 aspen carvings by Paco Chacon (photo opportunity courtesy of Dave Cogswell, Meeker, Colorado).

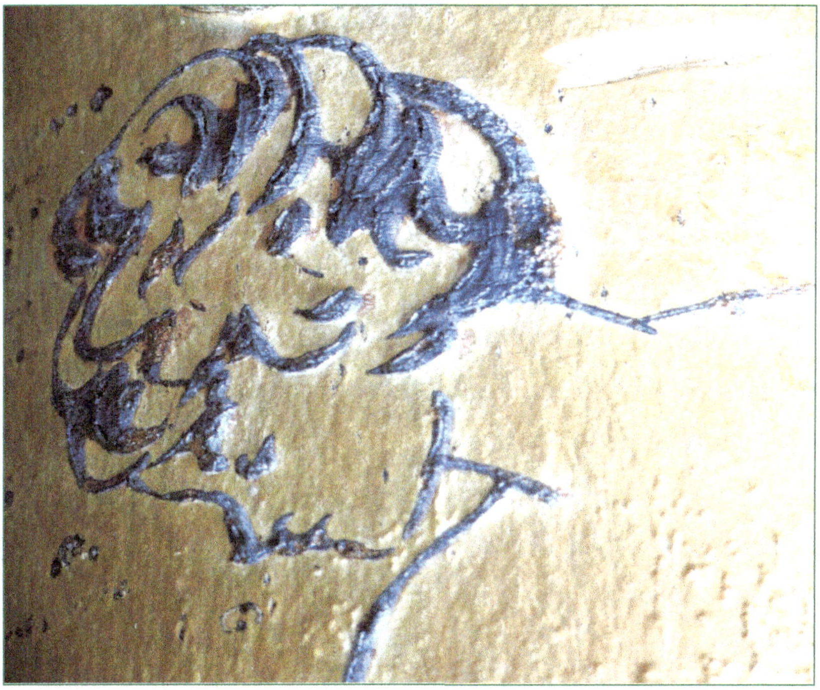

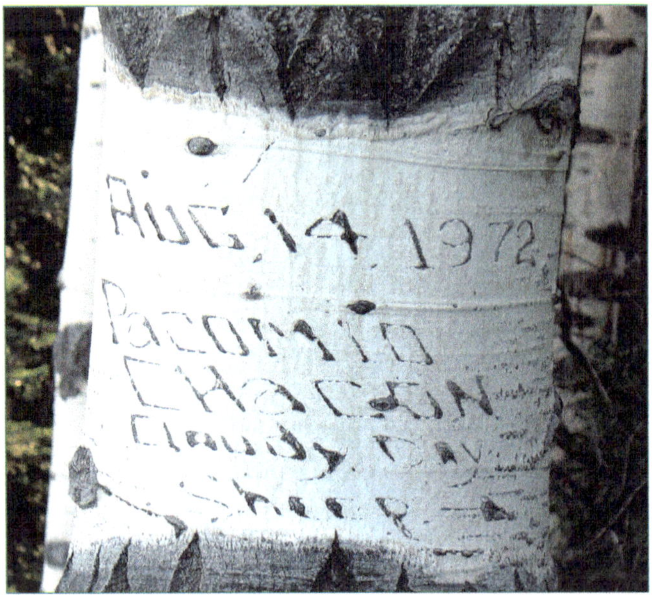

PAP 31: Paco's signature and comments. Paco seems to have never completed his commentary about the sheep in this image. He apparently was starting to say something about the sheep as he did in PAP 32.

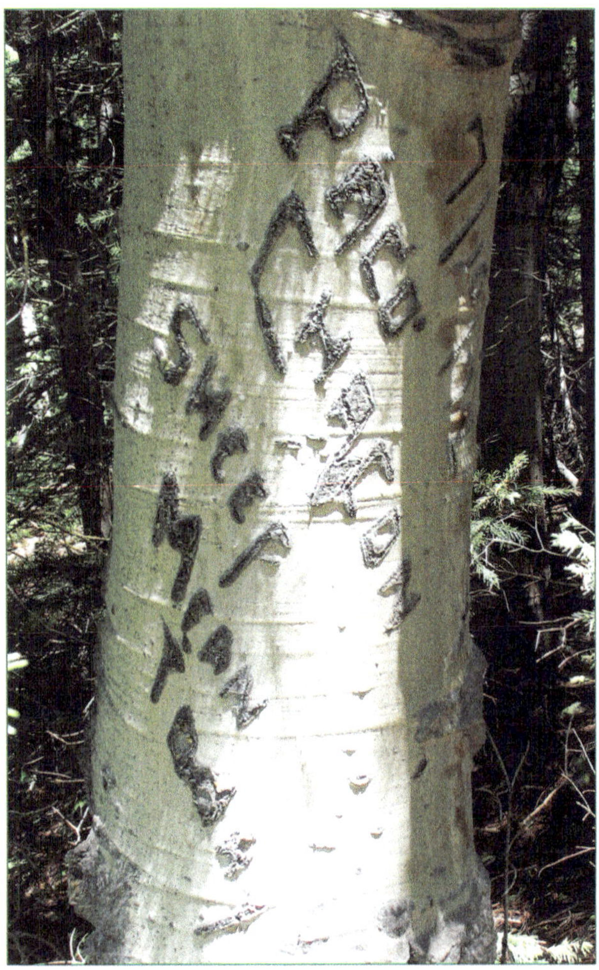

PAP 32: Paco's signature and comments on his sheep. This is clearly beginning to scar over into Stage 4 of the carvings life cycle.

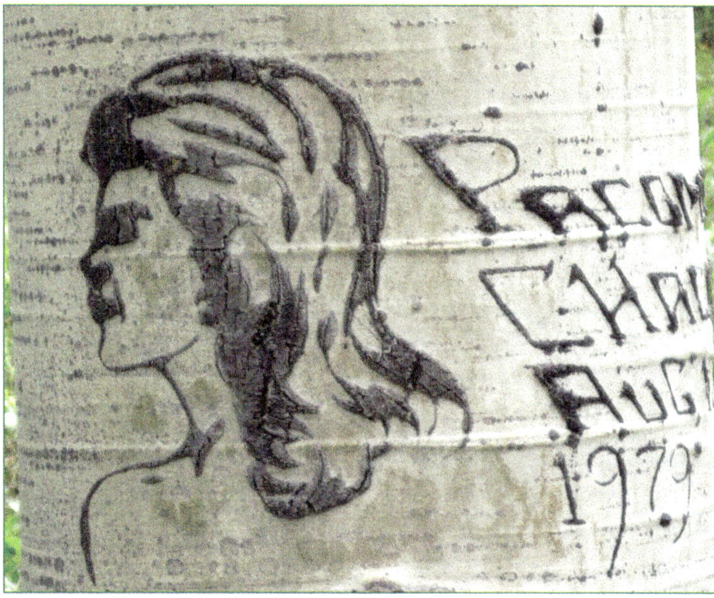

PAP 33: Paco's "Aspen Cameo."

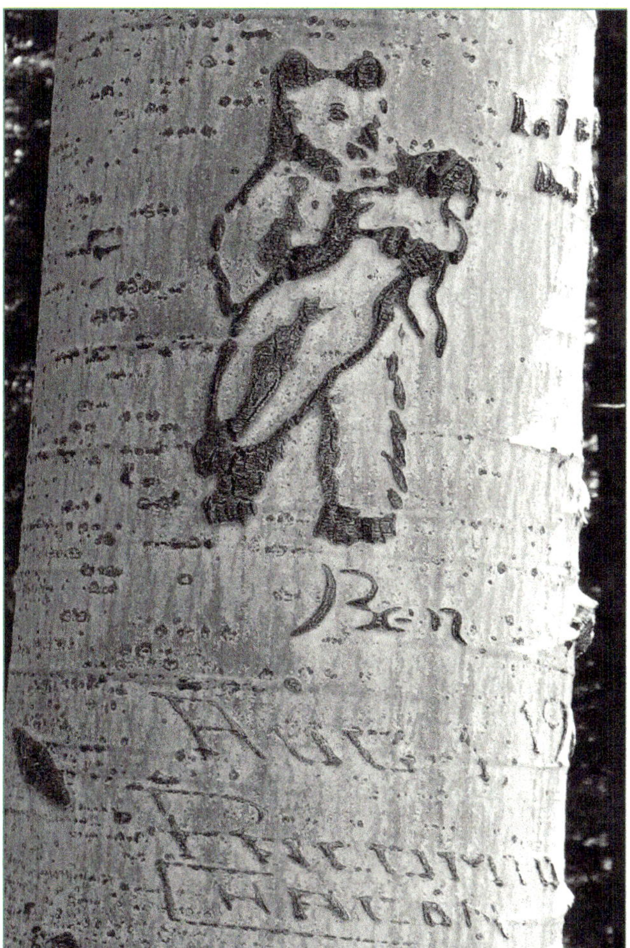

PAP 34: Paco's whimsical "Bear Stealing Woman" from the White River National Forest, Meeker, Colorado. Paco's dated signature on this tree is believed to post-date the carving of the bear with the woman.

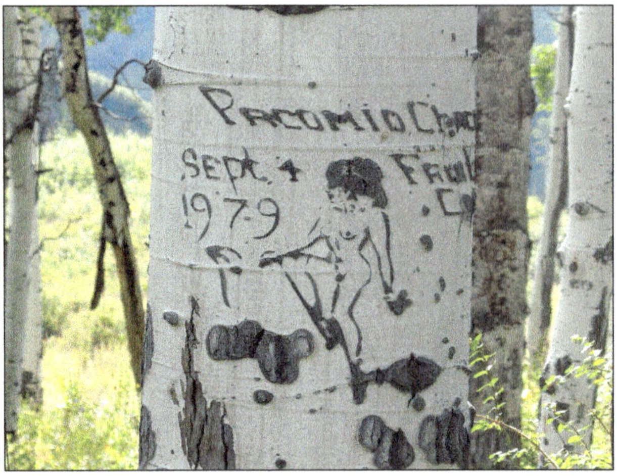

PAP 35: A typical Paco Chacon dendroglyph "calling card" with a nude woman, his signature, hometown, and date of the carving.

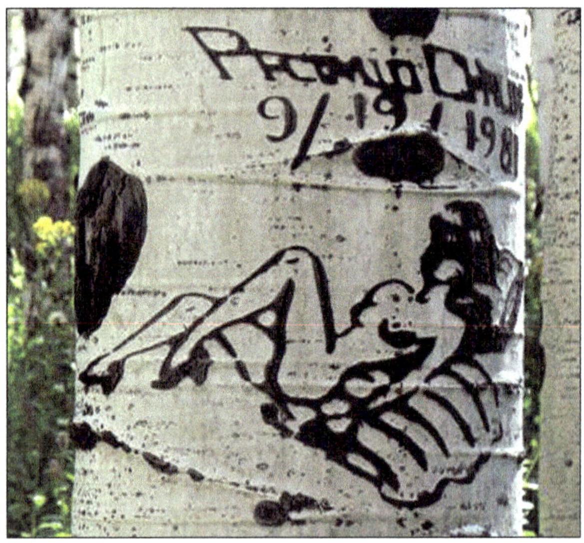

PAP 36: A typical Paco Chacon dendroglyph "calling card" with a nude woman ("Paco's Sunbather"), his signature, and date of the carving.

PAP 37: A typical Paco Chacon dendroglyph "calling card" with a nude woman ("Paco's Patience"), his signature, hometown, and date of the carving.

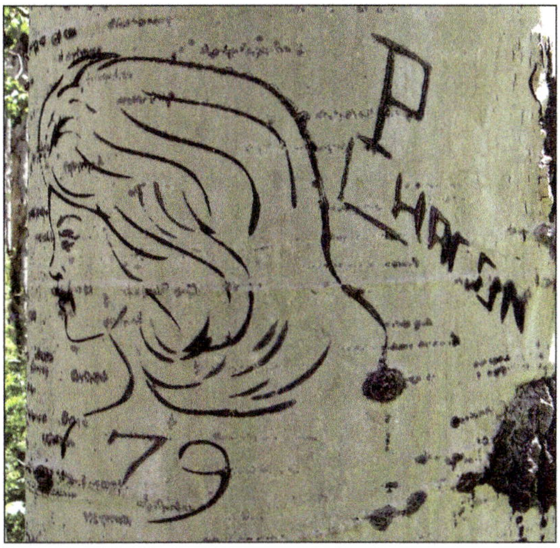

PAP 38, 39, 40: Miscellaneous Chacon aspen carvings. No. 36 may well be the work of Paco's brother, Press Chacon who also herded sheep in this area (no. 38 is from the Cogswell Collection, Meeker, Colorado).

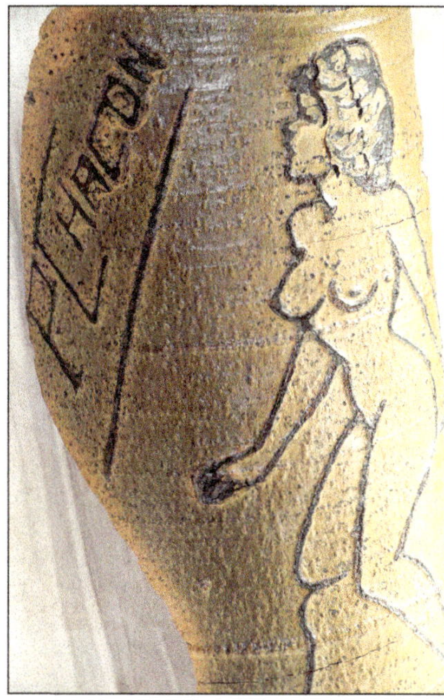

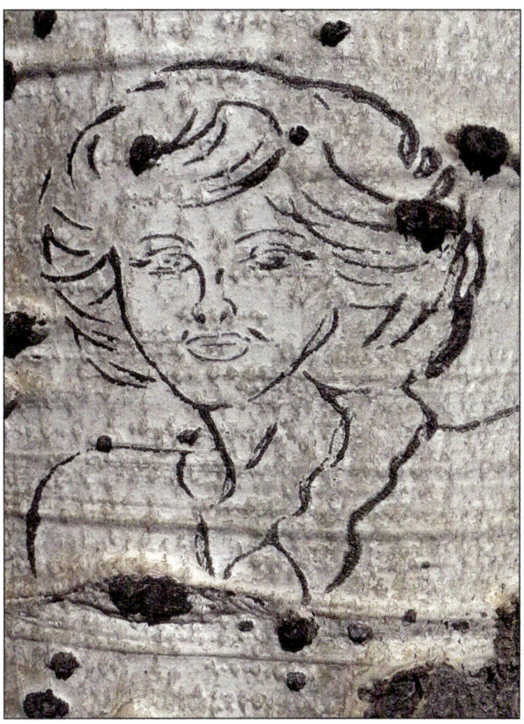

PAP40b: Paco's "Honey" courtesy of Troy Osborn, Meeker

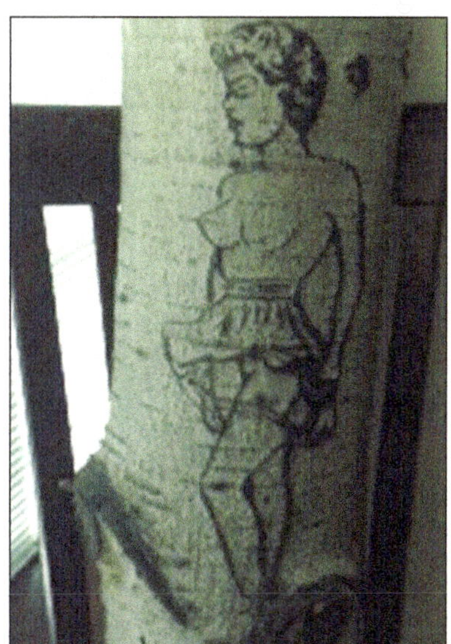

PAP 40a: Paco's "Bliss" courtesy of Peggy Halandras, Chicago, Illinois

PACO'S GROUP TWO ASPEN CARVINGS

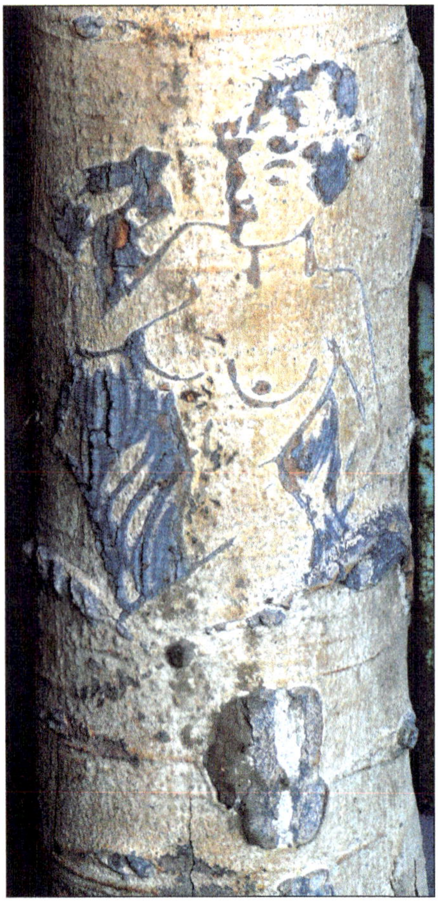

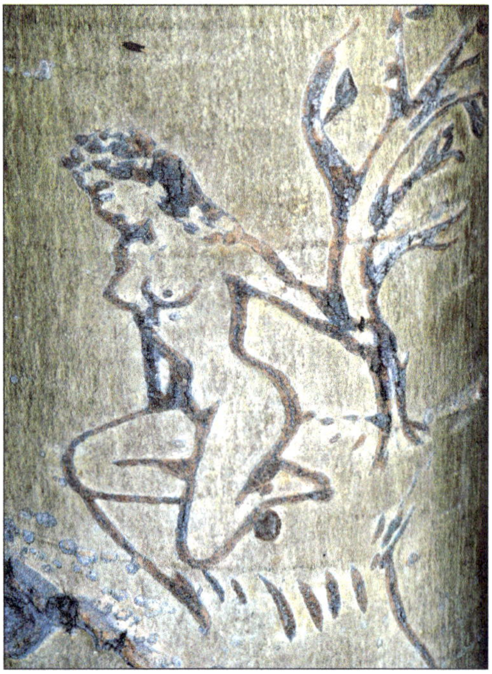

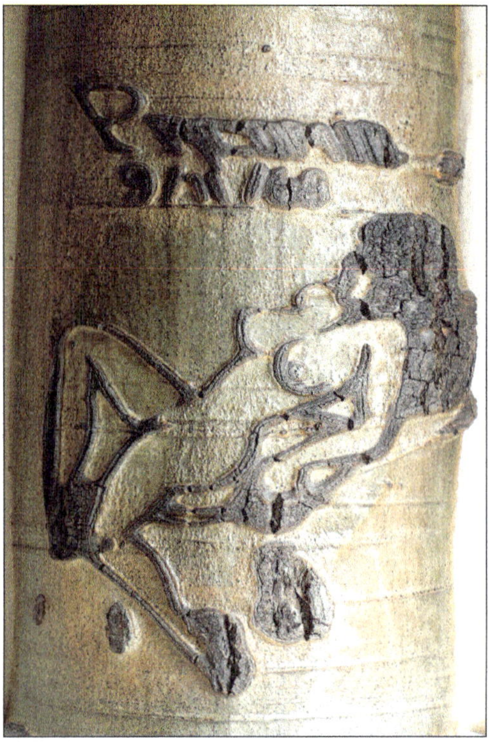

PAP 41, 42, 43 are typical Paco aspen "calling cards (photo opportunity courtesy of Dave Cogswell, Meeker, Colorado).

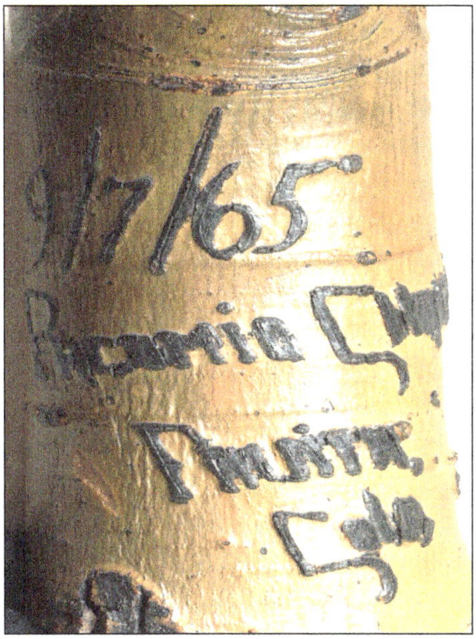

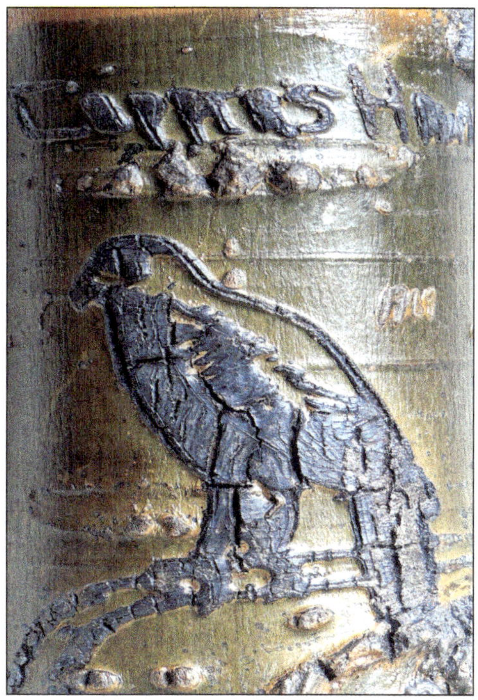

PAP 44 (signature), 45 (horse), 46 (Coopers Hawk) (photo opportunity courtesy of Dave Cogswell, Meeker, Colorado).

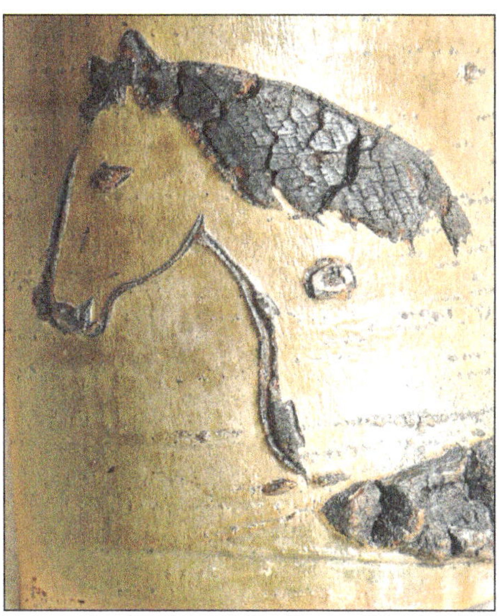

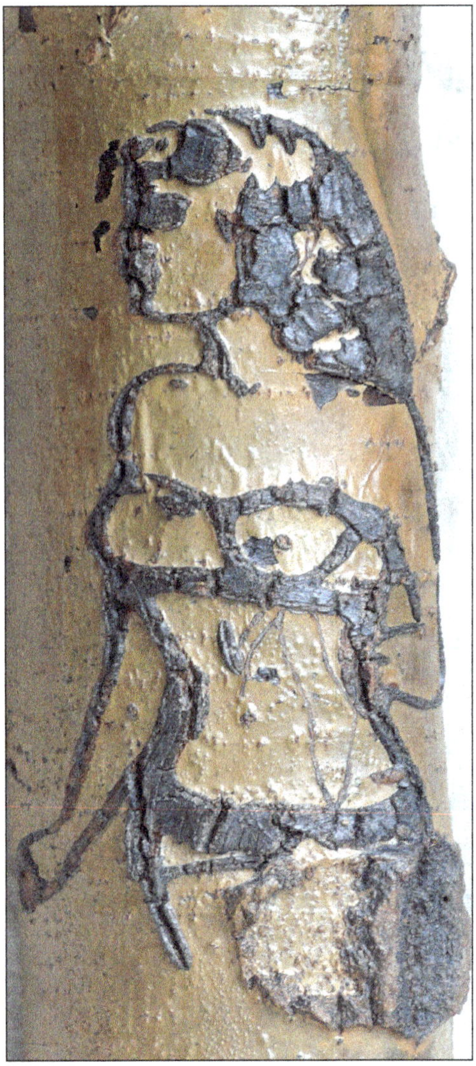

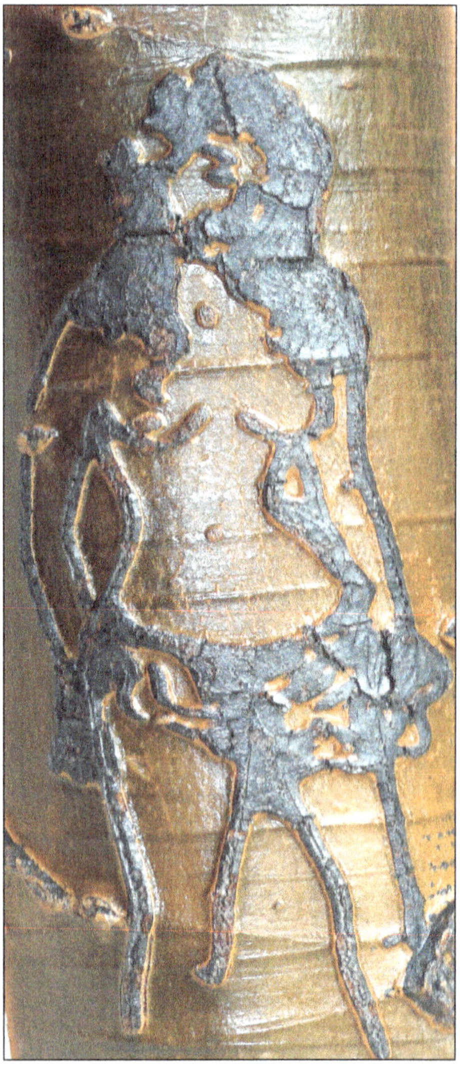

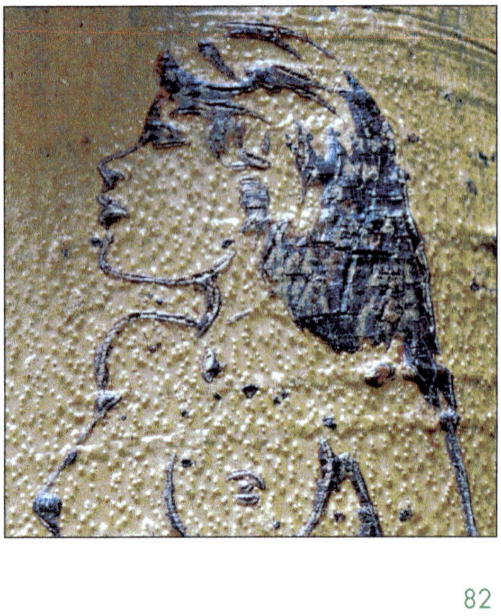

PAP 47, 48, 49 are typical Paco aspen "calling cards" (photo opportunity courtesy of Dave Cogswell, Meeker, Colorado).

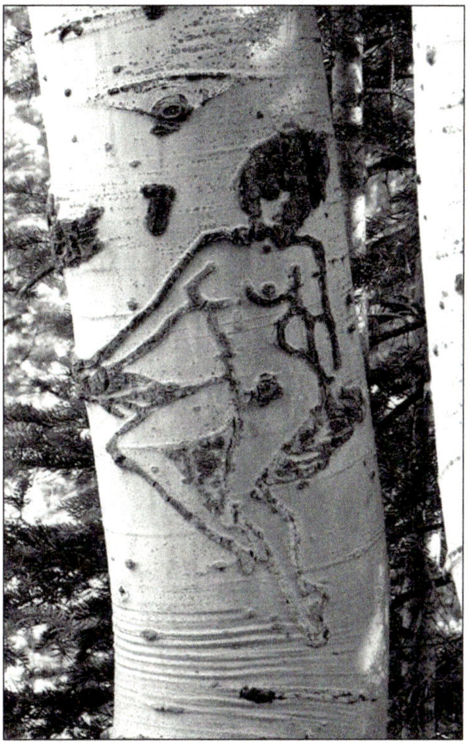

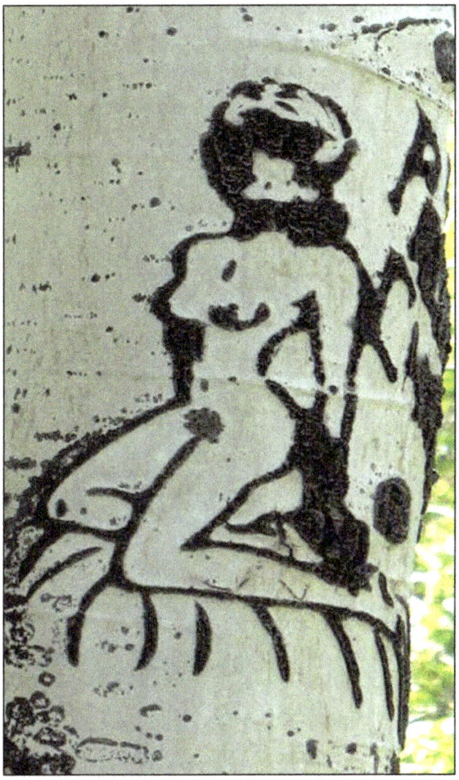

PAP 50, 51, 52. Typical Paco aspen "calling cards."

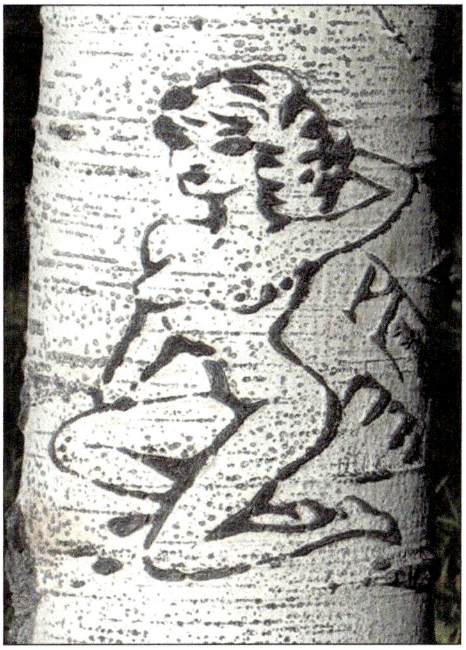

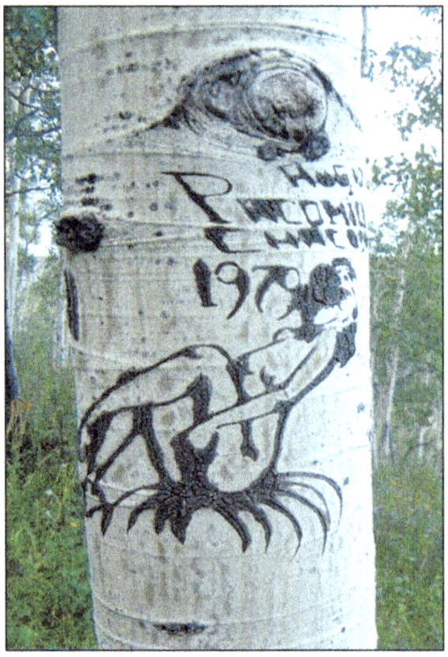

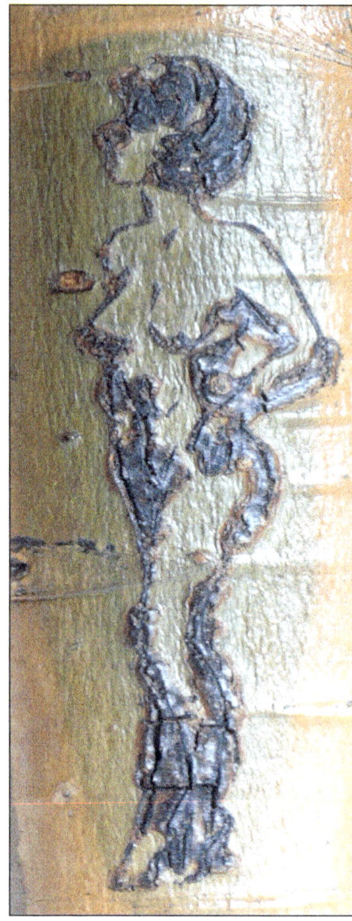

PAP 53, 54 (no. 54 is from the Cogswell Collection, Meeker, Colorado).

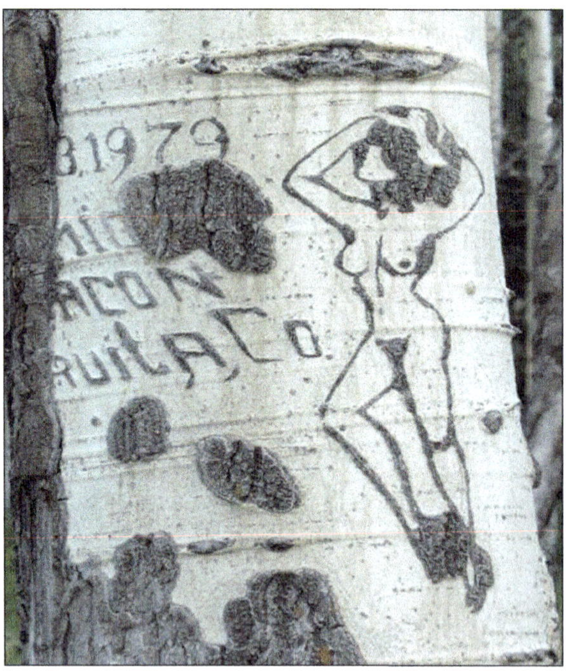

PAP 55: A 1979 Paco aspen "calling card."

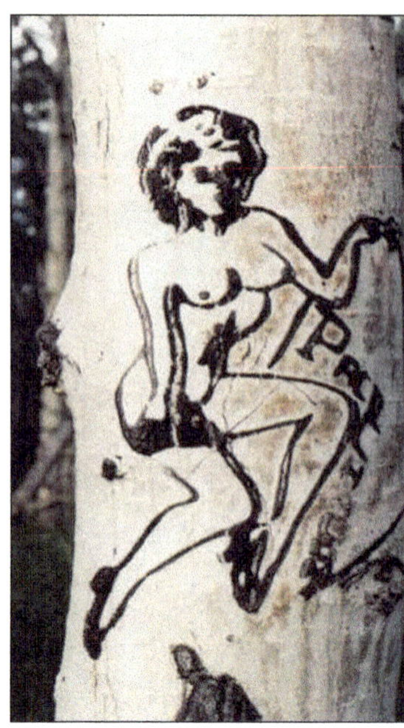

PAP 56: Typical Paco aspen "calling card."

Chapter 8

PACOMIO'S ROCK ART

Although he is known to have left an early (1939) signature on a cliff at Looking Glass Rock near Moab, Utah (Figure 7), Paco Chacon's rock art--his petroglyphs--is currently known to be found in two general areas. One is the region in and about Shavetail Basin on the Douglas Creek Arch roughly ten miles south of Rangely in Rio Blanco County, Colorado. This was the winter range for the flocks of Mike Theos of Meeker for whom Paco herded from 1972 until 1978. This range offered plentiful sandstone cliff faces of the Mesa Verde Formation that were well suited for carving (Figure 44).

Paco also worked for Louis Livingston of Craig, Colorado, from 1959-1972. Livingston apparently wintered his sheep near the Green River south of Vernal, Utah, where there were also sandstone cliffs suitable for carving. My only personal experience with Paco's rock art was in the region south of Rangely where he recorded a substantive portion of the examples illustrated herein. The areas where Paco worked in

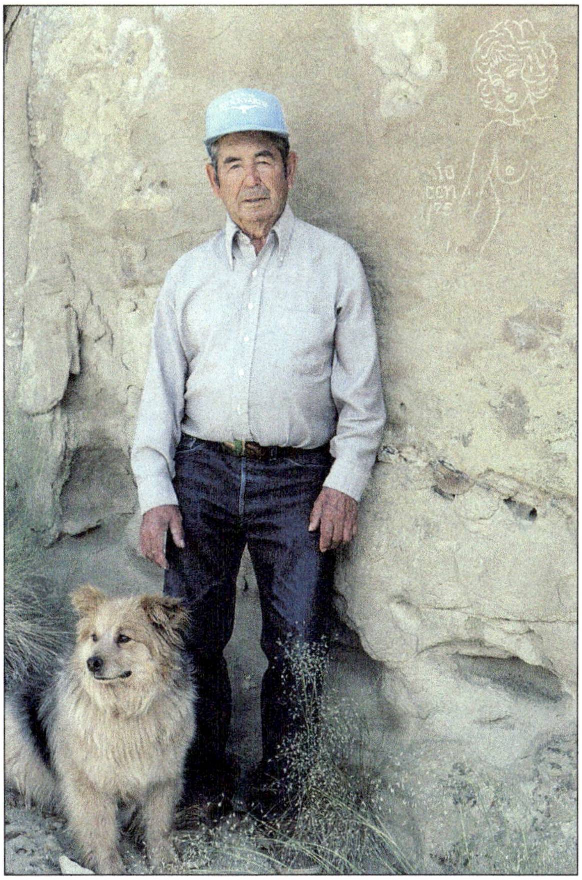

Figure 44: Paco Chacon in 1992 posing with his 1975 carving "Little Miss Locket" and my dog, Cinco de Mayo, in Shavetail Basin south of Rangely, Colorado. Paco was a true dog lover and a good dog trainer.

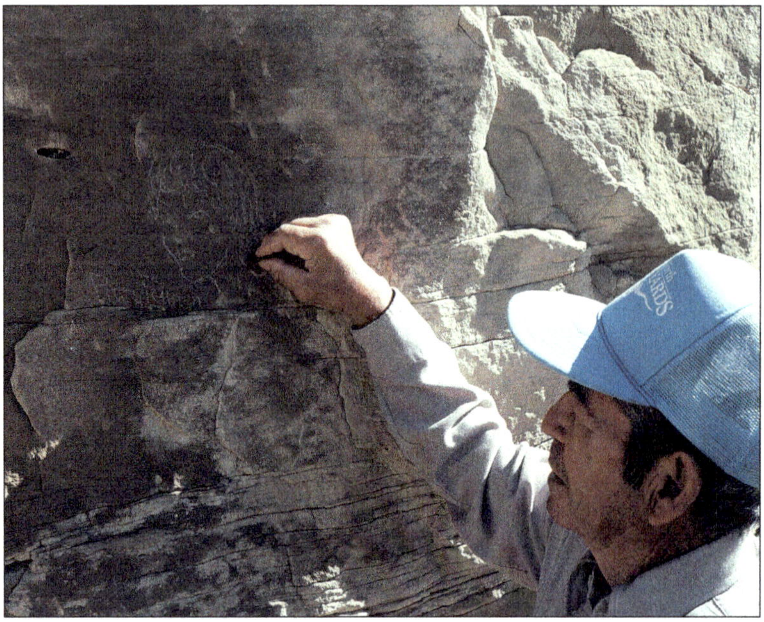

Figure 45: Paco Chacon demonstrating how he outlined a figure on a cliff face with a small rock before beginning to carve it. 1992 photo is from the Shavetail Basin area south of Rangely, Colorado.

Utah are largely on what is now part of the Northern Ute Indian Reservation where access is controlled. My colleagues and I have to date been totally unsuccessful in obtaining permission from the tribe to record Paco's works on the reservation.

These unrecorded works unfortunately include Paco's previously discussed image of the mountain lion attacking a lamb. There are thus only a few examples of Paco's rock art included here that are from Utah. Most of them have been provided by my colleague, Jim Truesdale, of Laramie, Wyoming, who has served as an archaeological consultant for the tribe. There may well be other areas out in the canyon country, such as his "Canyon Bride" in Salt Canyon near Canyon Lands National Park where Paco may have left some further yet undiscovered examples of his artwork.

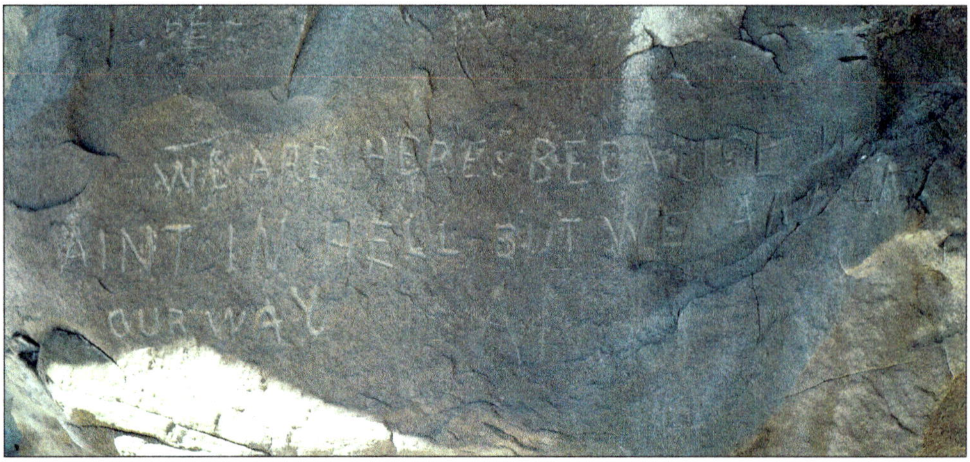

Figure 46: Sheepherder's or cowboy's inscription testifying to the often inhospitable nature of the region about Shavetail Basin south of Rangely, Colorado.

The area south of Rangely has over the years been both sheep and cattle country. Although Mike Theos always contended that the area was some of the best sheep wintering grounds in the United States, one old sheepherder or cowboy obviously did not think much of it when he left this message on the wall of a small side canyon. As shown in Figure 46, he stated that "we are here because we ain't in hell but we are on our way."

This area is a land of strangely eroded rock outcroppings known as hoodoos and beautiful small canyons. My field crews and I always thought it was a lovely place to work in the spring and fall. In summer it is a brutal land of heat, dryness, and black gnats that most people, in keeping with the above quote, tend to dread encountering. Like the horseflies that can drive horses to run through wire fences, black gnats can be so bothersome that they have been known to drive many, including one otherwise quiet and non-cursing archaeologist I know, namely Phil Born of Grand Junction, into a screaming and cursing frenzy characterized by drooling and running in mad circles. Fortunately, Paco did not have to deal with the gnats because he was on the summer ranges high in the mountains during their season.

The region offers beautiful, soft, and easily carved sandstone cliffs in multitudes. Paco readily took advantage of these and left a generous amount of carvings there. They were a true pleasure to discover and record. Although Paco's aspen art is certainly appealing, I have always admired his rock art the most, perhaps because it is so rare for herders to carve in the rock.

Paco's method of carving the cliffs was quite simple. As illustrated in Figure 45, he used a pencil, a small stone, or a nail to lightly scratch in the design he was envisioning. In his saddlebag he always carried an old screwdriver and a hammer. With these simple tools, he then began to laboriously chip or scratch out the basic design. He continued with the screwdriver to abrade and smooth out the outlines of his images. By use of various sized lines, he would then detail and texture the image. At one point he began to experiment with abrading techniques, apparently with rocks, to soften and add obvious texture to his petroglyph images. This is particularly evident in the treatment of the eyes of some of his figures in his portfolio.

PACOMIO'S ROCK ART PORTFOLIO (PRA)

As with his aspen carvings, the following portfolio of Paco's rock art images contains photographs taken by different people over the years with a variety of cameras and film. They are, therefore, also not at all consistent in their color or quality. I would again remind the reader that this was very much a "salvage project" intended to record examples of as much of Paco's work as possible before it was gone completely.

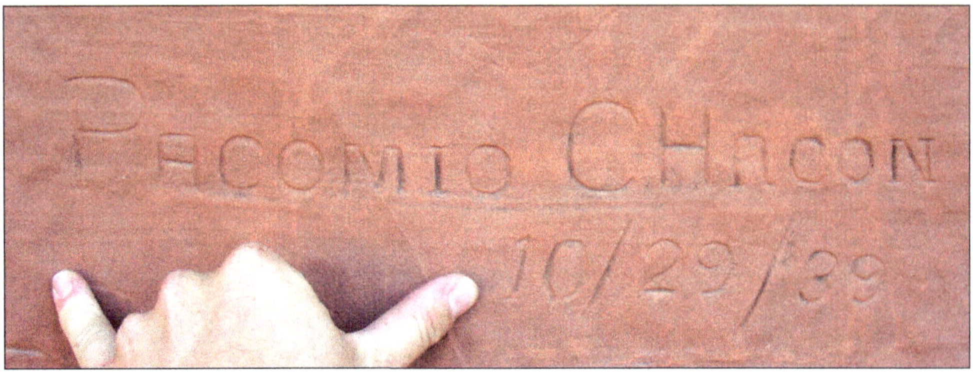

PRA 1: The earliest known rock art signature of Paco Chacon. This is located at Looking Glass Rock near Moab, Utah, and was carved when Paco was twenty-three years old and probably wintering sheep in the area for Steve Herndon of Norwood, Colorado (Table 1). This image has recently circulated some on the internet with people asking who Pacomio Chacon was (courtesy of Matthew S. Baker, Montrose, Colorado).

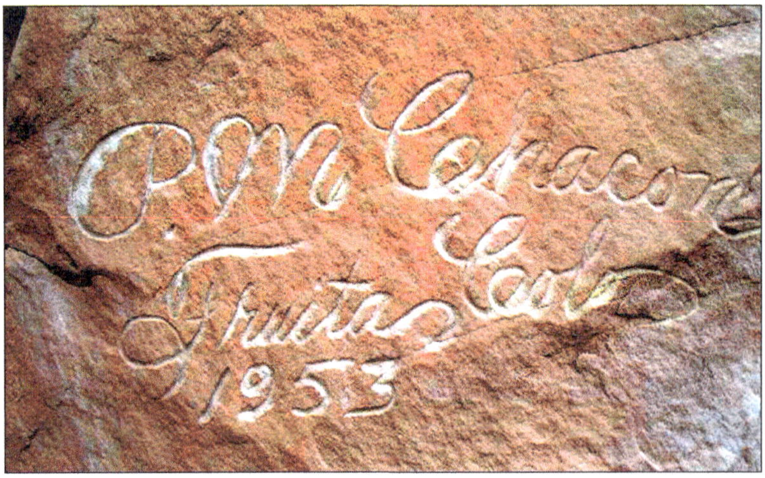

PRA 2: An early cursive rock art signature of Paco Chacon from Cottonwood Canyon, Utah. This was apparently carved while Paco was wintering sheep in Utah for Emmet Elizando of Montrose, Colorado (Table 1) (courtesy of Jim Truesdale, Laramie, Wyoming).

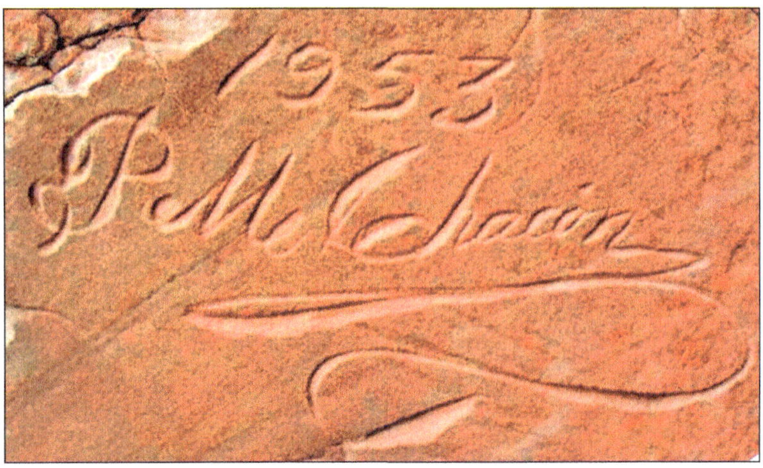

PRA 3: An early cursive rock art signature of Paco Chacon from Cottonwood Canyon, Utah. This was apparently also carved while Paco was wintering sheep in Utah for Emmet Elizando of Montrose, Colorado (Table 1) (courtesy of Jim Truesdale, Laramie, Wyoming).

PRA 4: An early and exquisite but unsigned work of Paco Chacon from Salt Canyon near Canyon Lands National Park in Utah. This carving ("Paco's Canyon Bride") shows just how good an artist Paco was with respect to his attention to detail. If one looks closely, the woman's left arm is subtly outlined behind the sheer bridal veil draping her arm. This obviously took considerable skill to envision and then carve so well. It is my belief that Paco had been planning to sign and date this carving on the round rock he outlined under the figure's left hip. This is one of Paco Chacon's finest rock carvings, and it probably dates from about 1953 while he was wintering sheep in Utah for Emmet Elizando of Montrose, Colorado (courtesy of John Ogden of Moab, Utah).

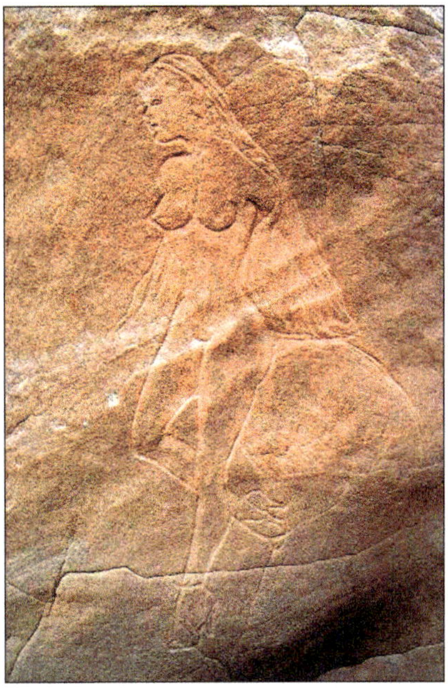

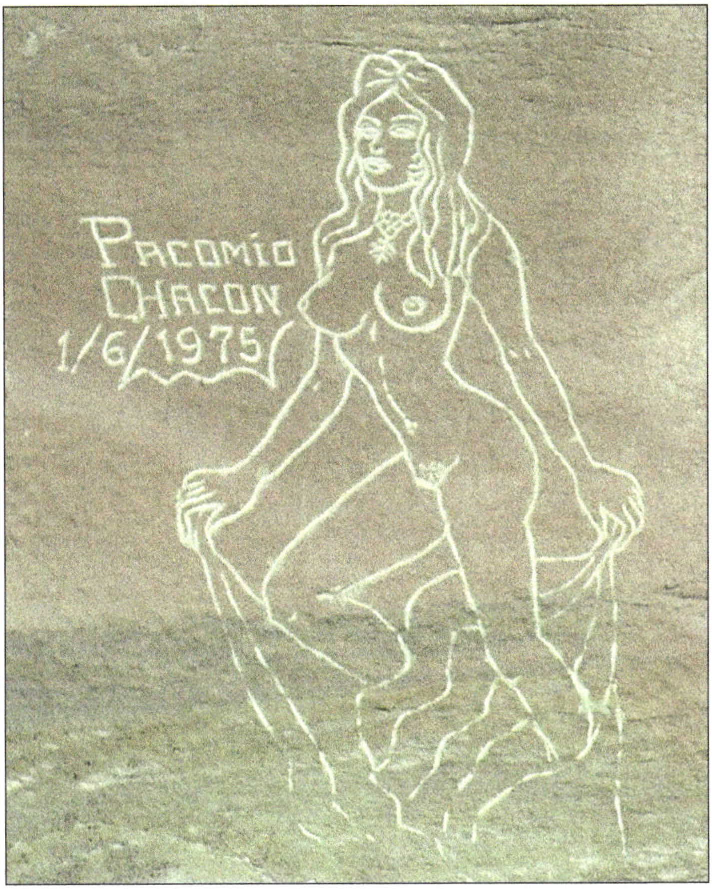

PRA 5: Paco's "Niñita Caprichosa" ("Little Whimsical Gal") from Shavetail Basin vicinity near Rangely, Colorado. This was carved while Paco was wintering sheep for Mike Theos of Meeker, Colorado. This is quite obviously among his finest works (author's collection).

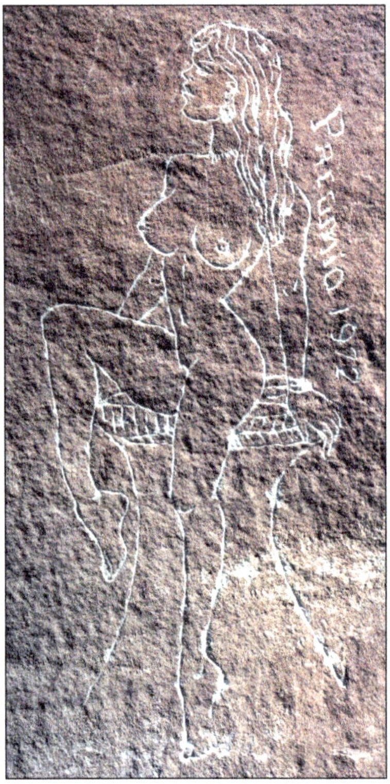

PRA 6: Paco's 1972 "Pedestal Lady" from Cottonwood Canyon near Vernal, Utah. Paco carved this while wintering sheep for Louis Livingston of Craig, Colorado (courtesy of Jim Truesdale, Laramie, Wyoming).

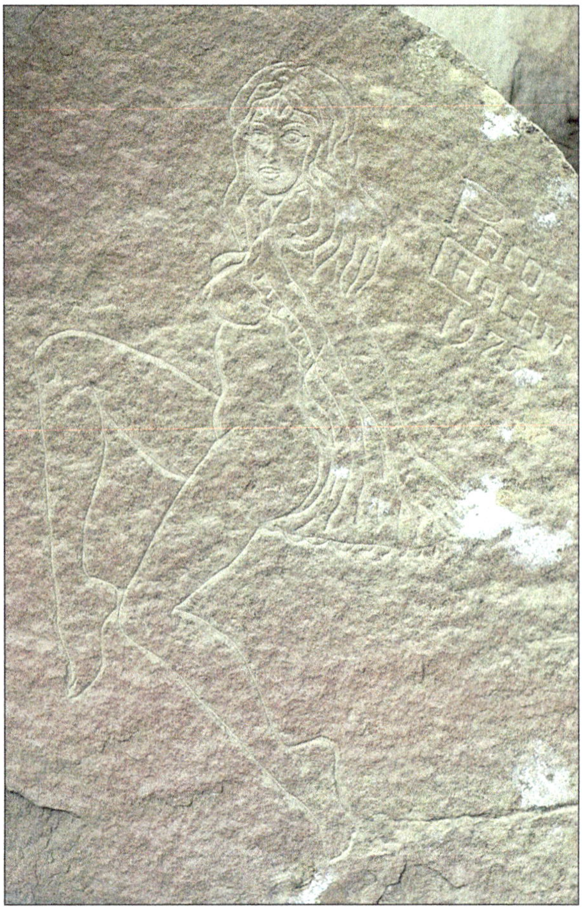

PRA 7: Paco's 1975 "Evening Breeze" from Shavetail Basin vicinity near Rangely, Colorado. This image is about one meter high and was carved while Paco was wintering sheep for Mike Theos of Meeker, Colorado. This is yet another example of his finest works (author's collection).

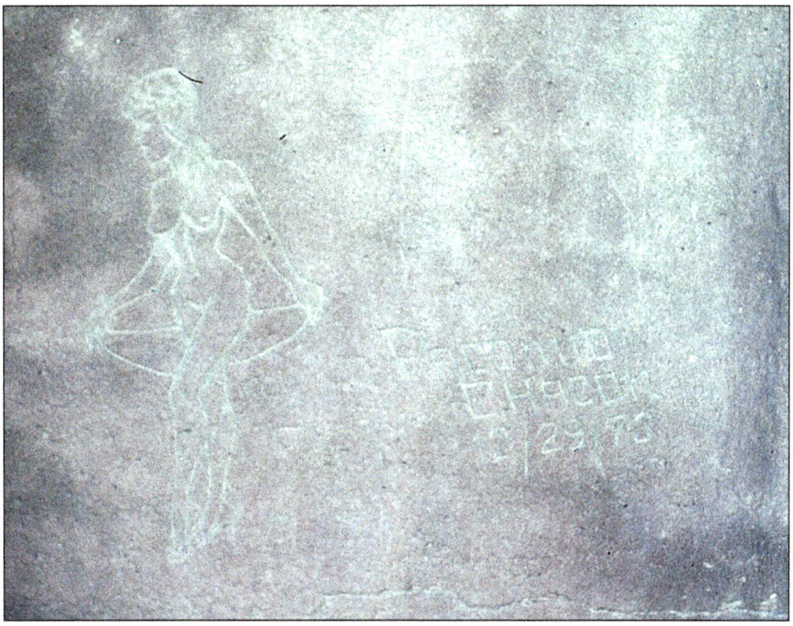

PRA 8: Paco's 1973 "Lady a Toweling" from Shavetail Basin vicinity near Rangely, Colorado. This was carved while Paco was wintering sheep for Mike Theos of Meeker, Colorado (author's collection).

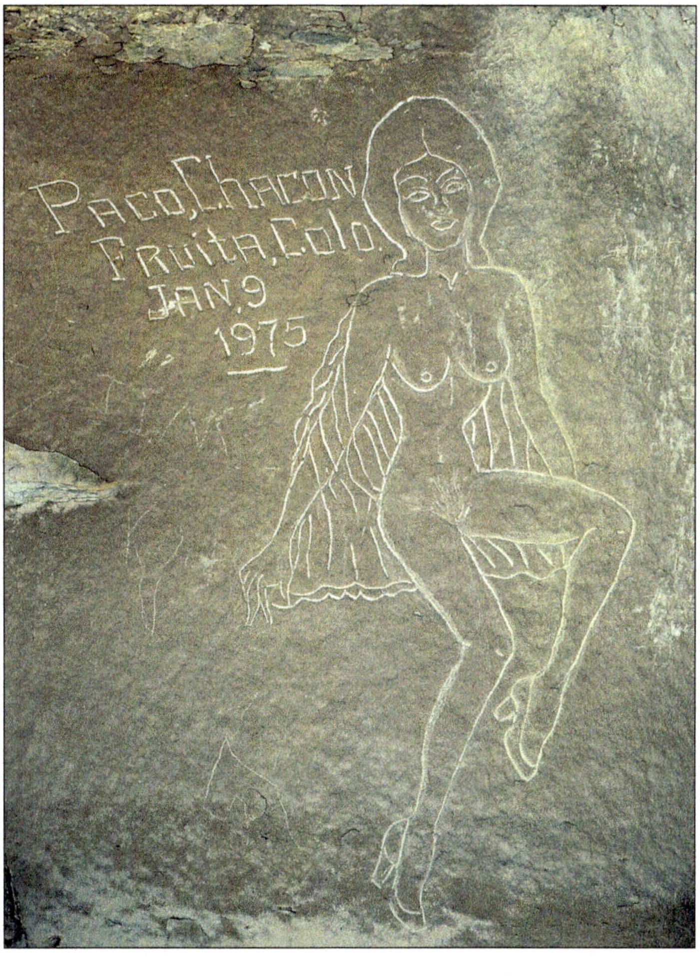

PRA 9: Paco's 1975 "Dream in Gossamer" from Shavetail Basin vicinity near Rangely, Colorado. This was carved while Paco was wintering sheep for Mike Theos of Meeker, Colorado (author's collection).

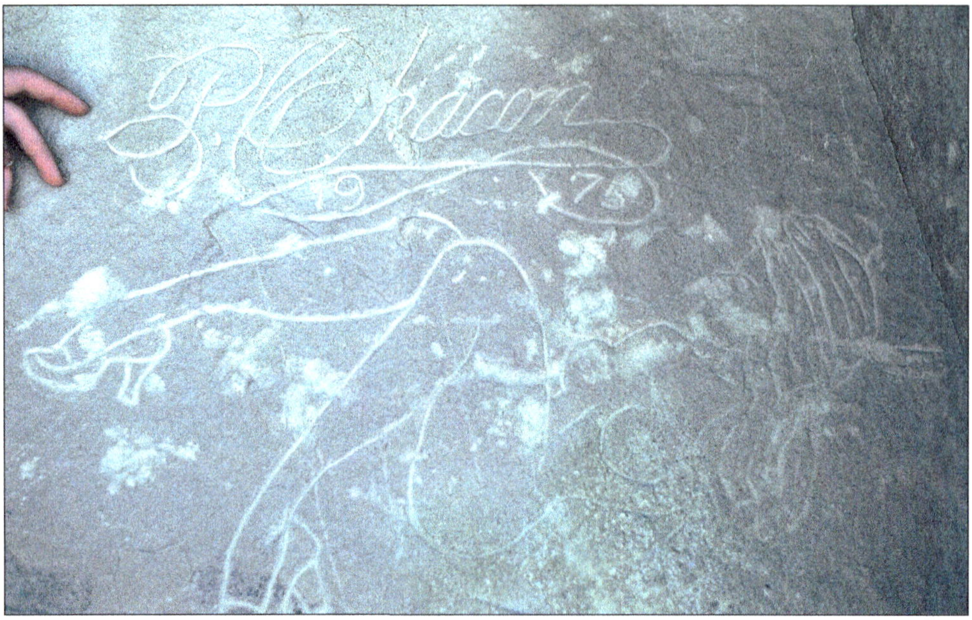

PRA 10: Paco's 1975 " Native Treat" from Shavetail Basin vicinity near Rangely, Colorado. Although difficult to see, Paco put an eagle feather in this lady's Indian-style headband (author's collection).

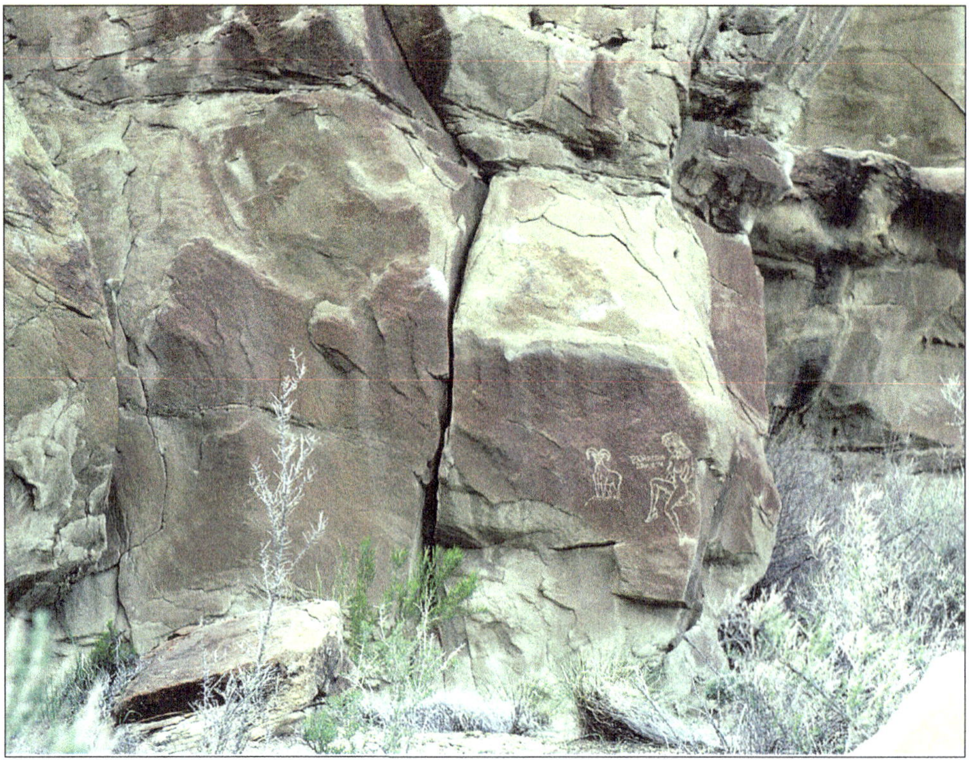

PRA 11: Typical context in which Paco's rock art is found in the Shavetail Basin vicinity near Rangely (author's collection).

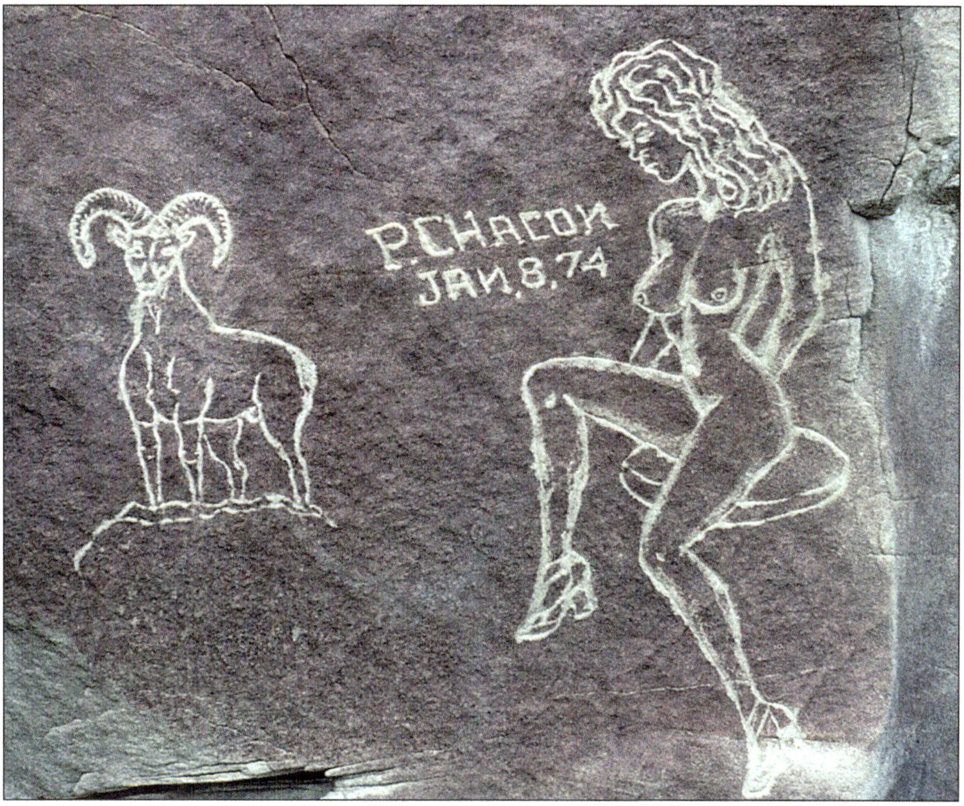

PRA 12: Details of carvings in image PRA 11. A Desert Big Horn sheep and Paco's "Little Miss Tuffet" (author's collection).

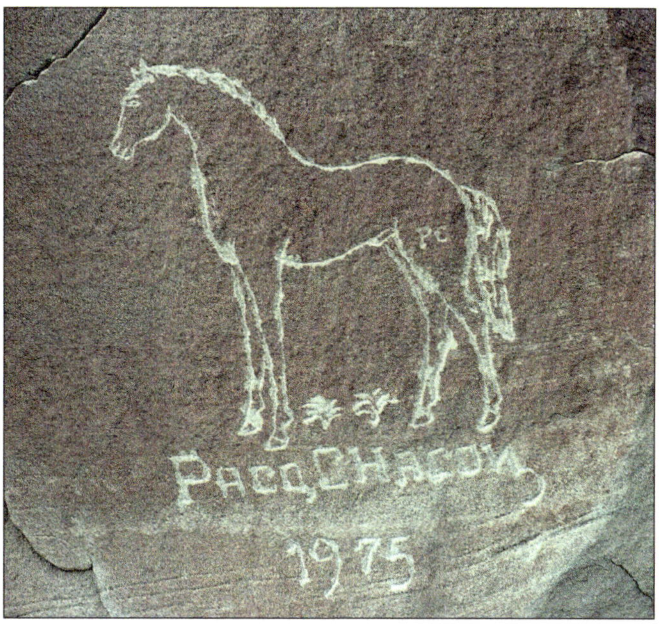

PRA 13: "Paco's Pony" from Shavetail Basin vicinity near Rangely, Colorado (author's collection).

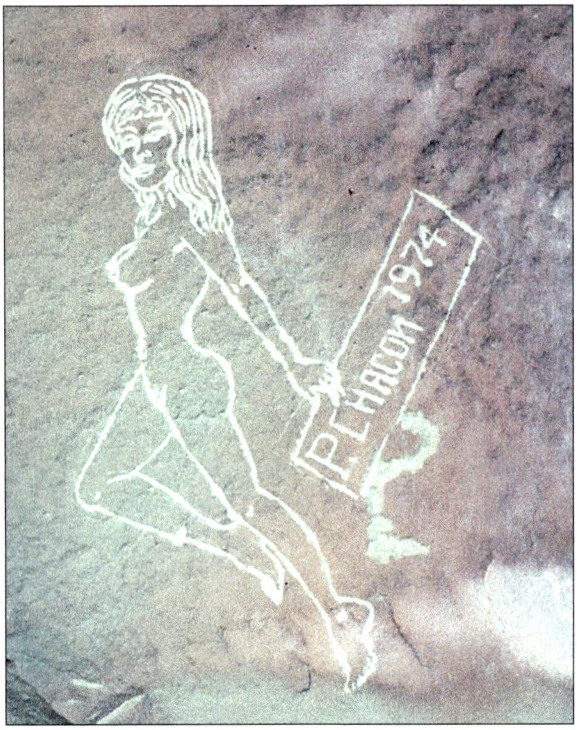

PRA 14: Paco's "Little Advertiser" from Shavetail Basin vicinity near Rangely, Colorado (author's collection).

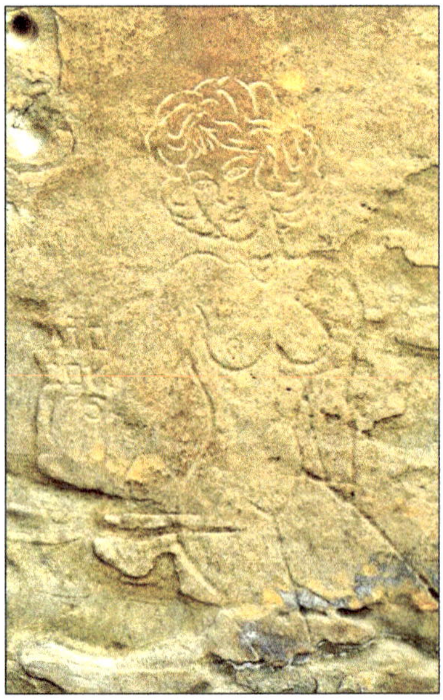

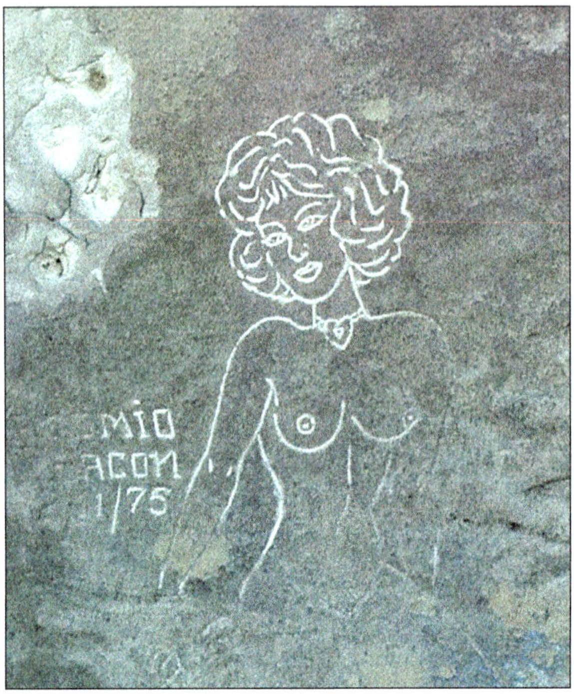

PRA 15a & b: Two images of Paco's "Little Miss Locket" from Shavetail Basin vicinity. When last photographed only a few years after she was first photographed ca. 1990 (above) she was rapidly deteriorating (author's collection).

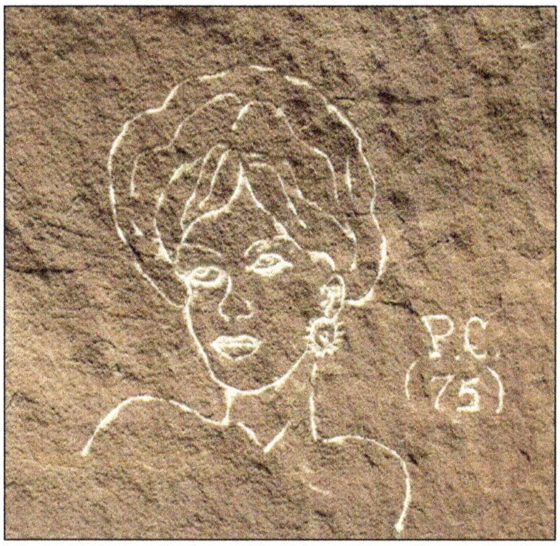

PRA 16: Paco's "Diamond" from Shavetail Basin vicinity near Rangely, Colorado (author's collection).

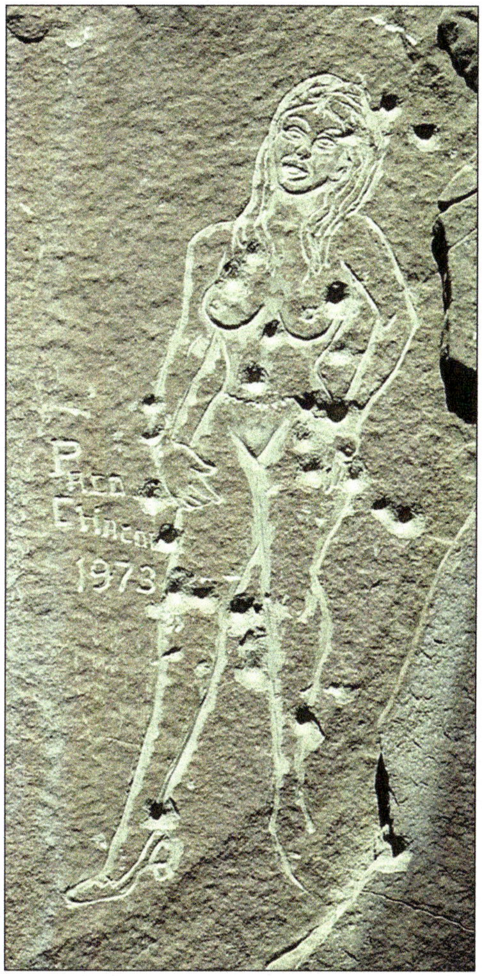

PRA 17: Paco's "Murdered Lady" from Shavetail Basin vicinity near Rangely, Colorado, as photographed in 1989 (author's collection).

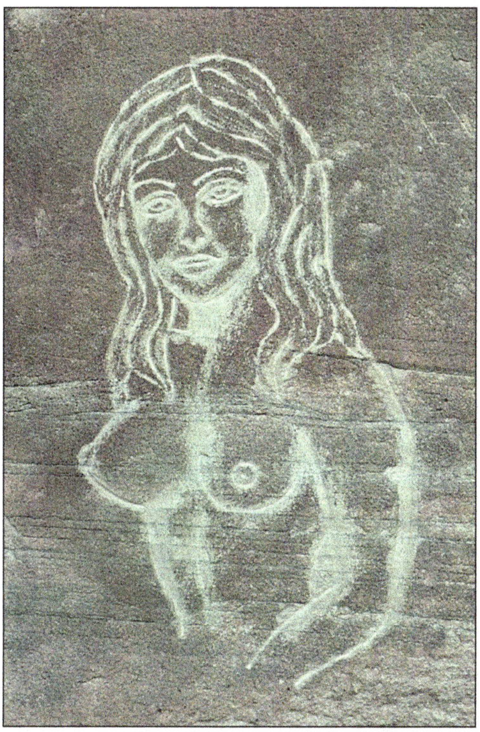

PRA 18: Paco's "Morning Eyes Awaken" from Shavetail Basin vicinity. In this study Paco has obviously been experimenting with heavier techniques of abrading, seemingly by rubbing selected portions of the image with a rock (author's collection).

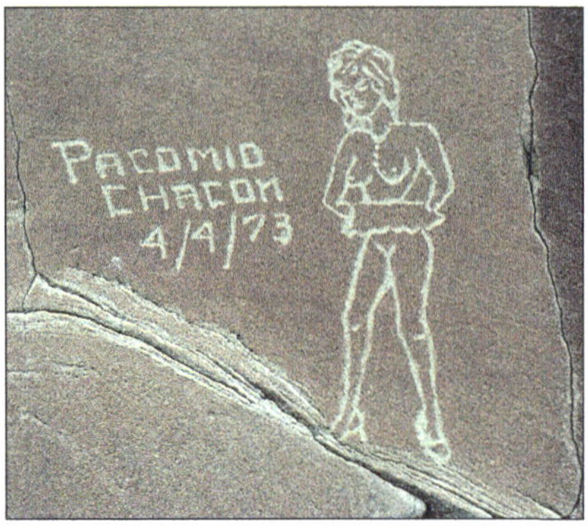

PRA 19: Paco's " Aproned Maid" from Shavetail Basin vicinity near Rangely, Colorado (author's collection).

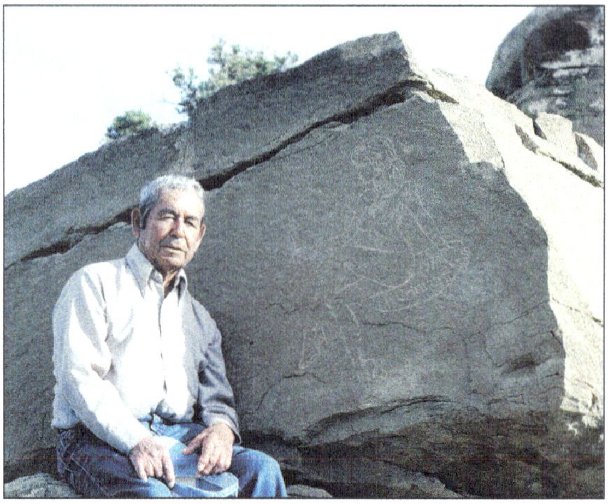

PRA 20: Paco with his "Blossom" from Shavetail Basin vicinity near Rangely, Colorado, ca. 1992 (author's collection).

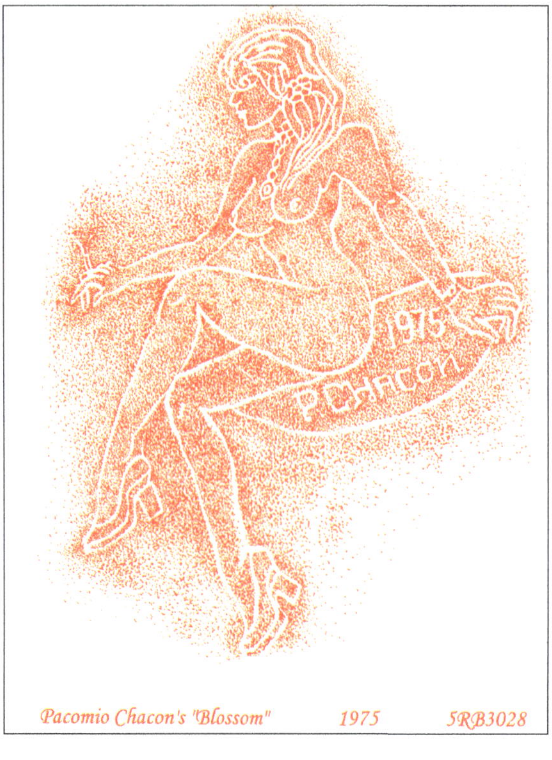

PRA 21: Artist's stippled image of "Blossom." Note the flower blossom in her hair (author's collection).

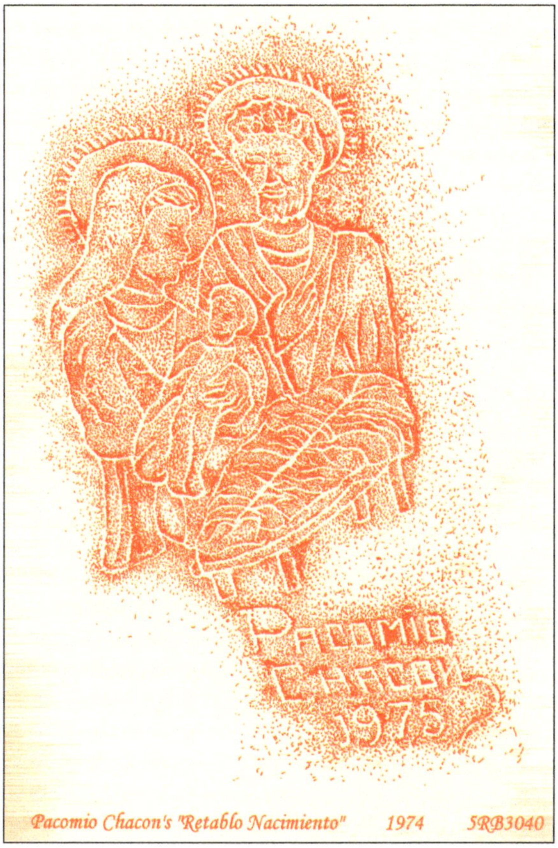

PRA 24: Artist's stippled image of Paco's "Retablo Nacimiento" (author's collection).

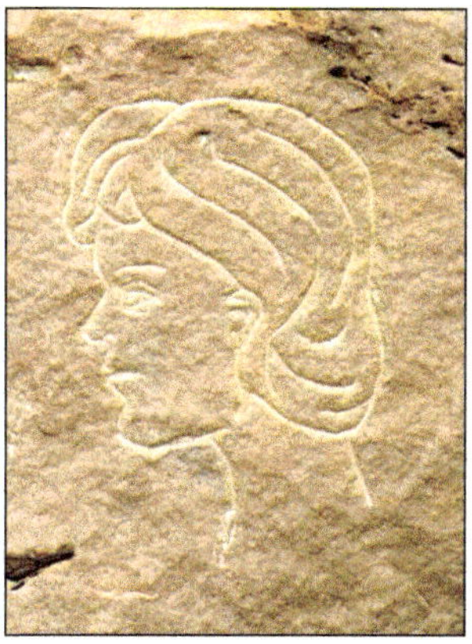

PRA 22: Paco's "Study in Cameo" from Shavetail Basin vicinity near Rangely, Colorado (author's collection).

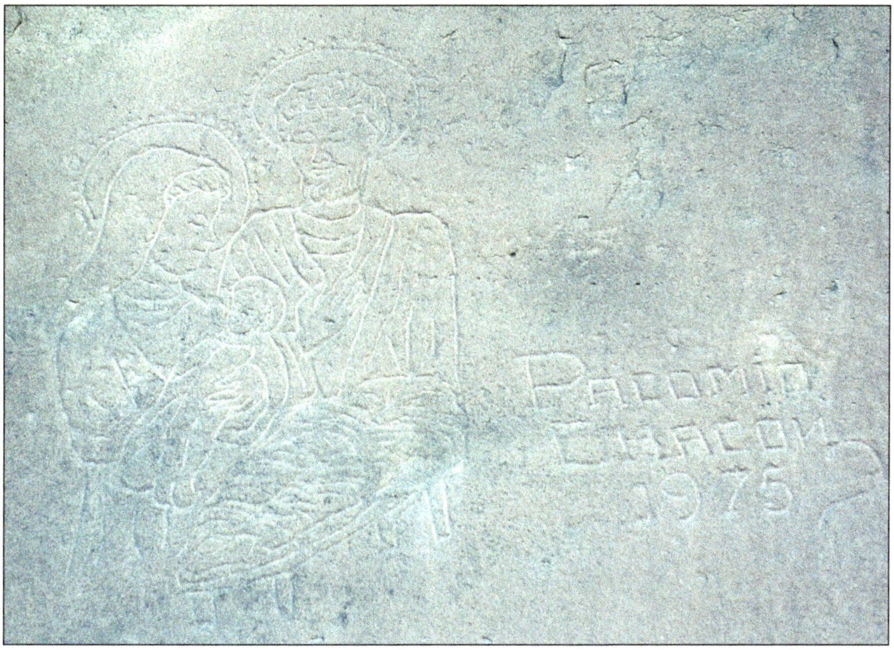

PRA 23: Paco's "Retablo Nacimiento" (Nativity Scene)from Shavetail Basin vicinity near Rangely, Colorado (author's collection).

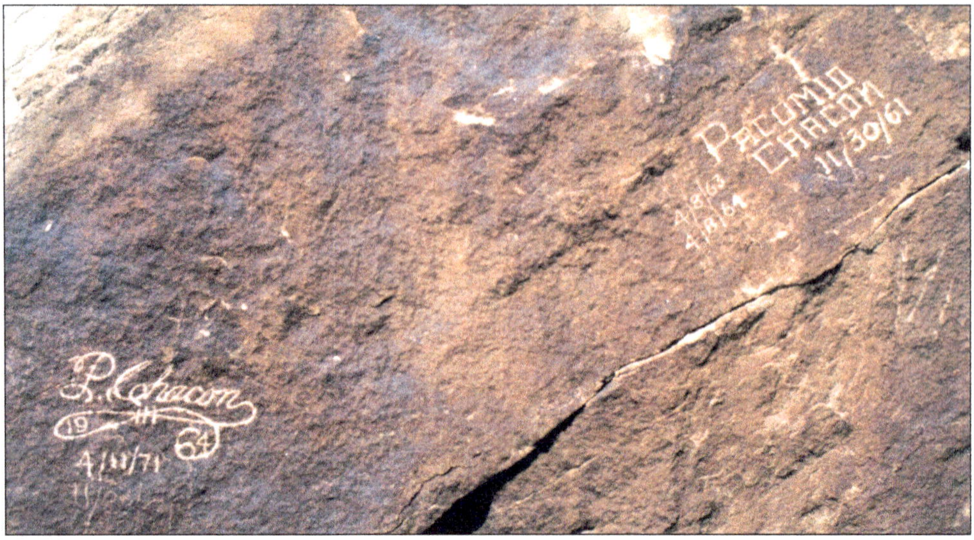

PRA 25: Paco's classic cursive and printed signatures from Natural Buttes Area of Uinta Basin near Ouray, Utah. These signatures were carved when Paco wintered sheep for Louis Livingston of Craig, Colorado (courtesy of Jim Truesdale, Laramie, Wyoming).

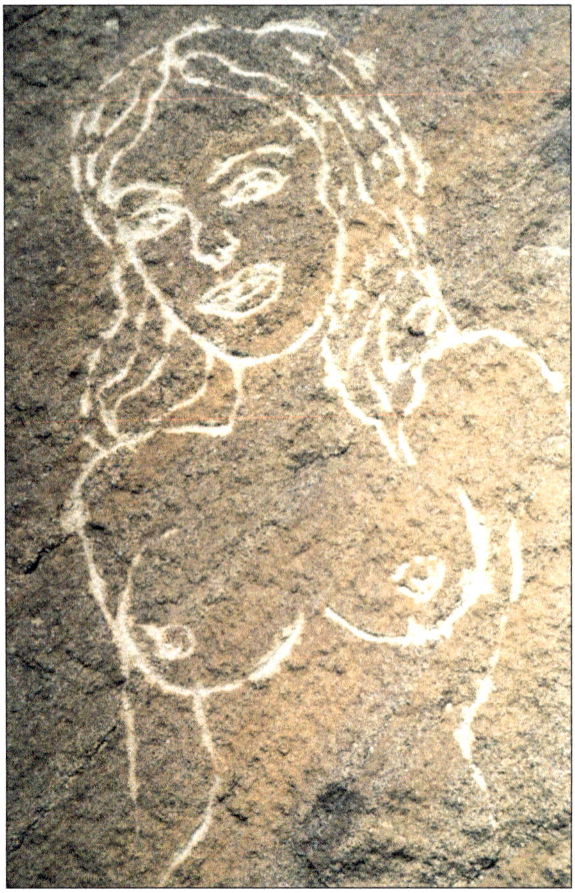

PRA 26: Paco's "Little Miss in Stone" near Ouray, Utah. This image was carved when Paco wintered sheep for Louis Livingston of Craig, Colorado (courtesy of Jim Truesdale, Laramie, Wyoming).

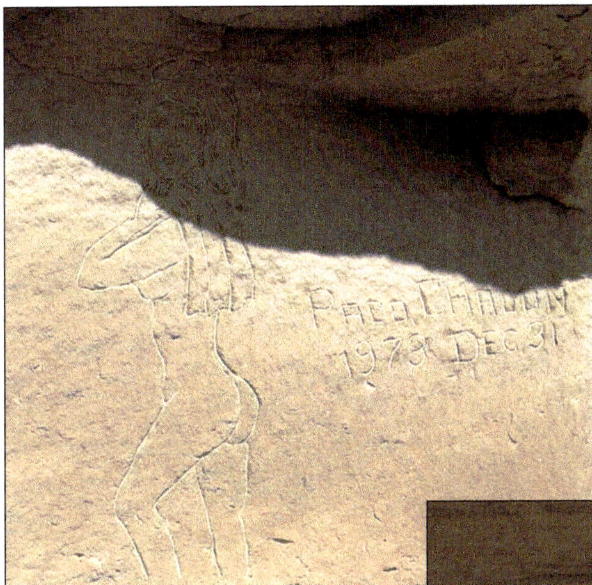

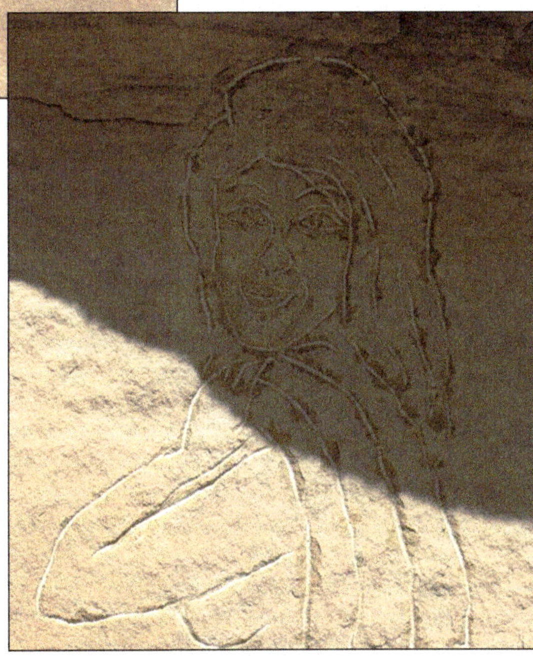

PRA 27a & b: Paco's "Lady with Towel" from Shavetail Basin vicinity near Rangely, Colorado (author's collection).

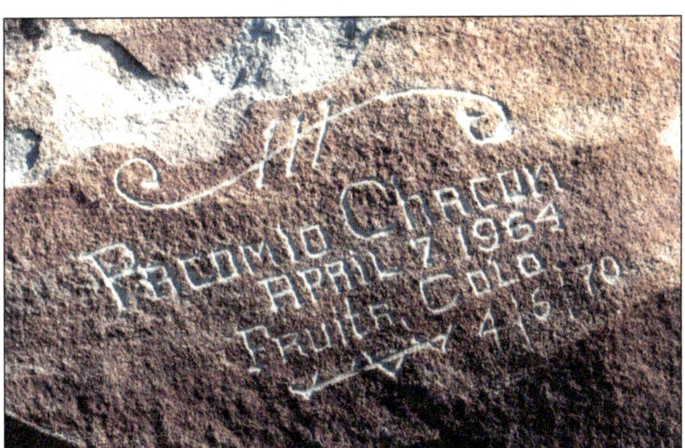

PRA 28: Elaborate block-printed signature of Pacomio Chacon from the Chapita Wells Gas Field in the Uinta Basin southeast of Ouray, Utah. This was carved when Paco was wintering sheep for Louis Livingston of Craig, Colorado (courtesy of Jim Truesdale, Laramie, Wyoming).

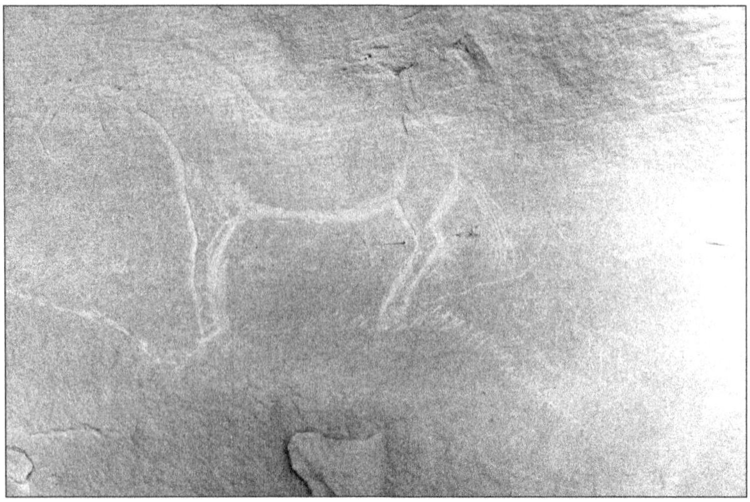

PRA 29: An image of a horse from the "Dueling Sheepherders' Cave," which was carved by someone with initials of "JPB" using techniques he had seemingly observed in Chacon's local work. It is clearly not of the quality of Paco's. This image was accompanied by a very poorly carved inscription reading "End of the Trail" (author's collection).

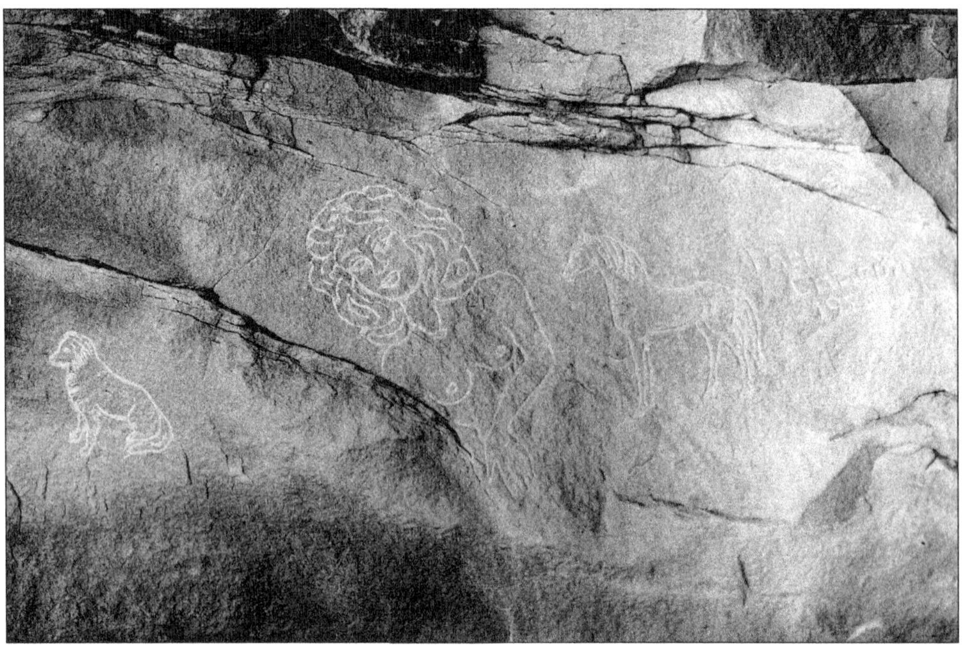

PRA 30: Once Paco saw the image and inscription in PRA 29, he decided he would show just how much better his work could be and drew these images of his dog "Kino," a bare breasted woman, and his horse. It was out of this circumstance that the site was named the "Dueling Sheepherders' Cave." Although it may not have happened just this way, it still makes for a good story and appropriate way to close this portfolio of Paco Chacon's rock art (author's collection).

Chapter 9

PACOMIO'S MISCELLANEOUS ART

Paco was compulsive about all of his artwork. If he did not have an aspen tree or cliff face handy, he picked up a pencil, pen, or crayon and started to doodle on anything available. Any scrap of paper or cardboard became his next canvas and he commonly incorporated his art into letters, grocery lists, and other everyday ephemeral, paper items. These works typically did not survive the ravages of time and we have no clue at all about what he might have drawn on paper while in his sheep camp. His subject range was obviously broad as indicated in the few salvaged images included here. He very commonly gave away his drawings, particular those he made of or on behalf of friends and family. As explained earlier, Paco's handwriting was so good that his Army buddies in World War II often commissioned him to write or at least address their letters home to family and sweethearts. He often included a colored picture of birds or flowers on the envelope.

It seems fair to suggest that Paco may have had a strong obsessive-compulsive streak embedded in his personality. His hands and mind did not know how to be idle, and he would often draw while still carrying on a decent conversation. There were likely few surfaces that Paco considered sacred. Although he was not a graffiti artist who defaced things other than the stock driveway signs included here, I can only imagine what kind of wonderful images he might have created on the side of a railroad car with a few cans of spray paint!

Toula Theos, the former wife of his 1970s employer, Mike Theos, once described how Paco provided them with the most beautifully written grocery lists. He prepared them in his exquisite calligraphy and then dressed them up with drawings of animals. Naturally, many bore the ever-present, whimsical little calendar girls that were his trademark. Toula lamented that she had not bothered to save any of Paco's grocery lists. Sadly, there are no surviving examples to include here. When out with the sheep Paco wrote home to his family and particularly to his children. His letters were in his beautiful script and adorned with all manner of little drawings of birds or animals. A few surviving examples of his drawings of this kind are included here.

Most of Paco's drawings and paintings in no way matched the quality of his carving in the mediums of the aspen and rocky cliffs. They indicate the limitations of his abilities as an artist and suggest that he never mastered many of the standard artistic techniques. Those skills, such as perspective, seem to have eluded him, and one does not have to be an art critic to see the difference in quality between his drawings and his cliff and aspen carvings, the techniques for which he obviously mastered. At one point late in his life Paco experimented with relief carving by attempting to carve a two dimensional head of an Indian woman with long braids on a block of sandstone behind his home. Unfortunately, this piece disappeared even though I did see it once. Paco also carved the name and birth and death dates of his former employer, Mike Theos, on a boulder at his grave site at the Theos Ranch near Yellow Jacket Pass north of Meeker. I have no photos of these two examples of his miscellaneous work but I did obtain one of the grave stone he made for his favorite dog, Kino, as shown here in PMAP 16.

A collection of Paco's drawings and miscellaneous artwork is included here to round out his portfolio and provide a comprehensive perspective on Paco as a folk artist and on his limitations. Although it is not held out as particularly good artwork, it is his and for an old sheepherder with only a few years of grade school and absolutely no art training, some of it is not so bad. As primitive folk art, some of this part of Paco's portfolio is actually pretty decent. Paco always tried to give me most of the examples of his drawings reproduced here. I never felt that I could accept them and left them with Paco. Unfortunately, all of them disappeared and I have no idea where they now may be. Paco, particularly when drinking, would commonly give his artwork away to both friends and strangers.

PACOMIO'S MISCELLANEOUS ART PORTFOLIO (PMAP)

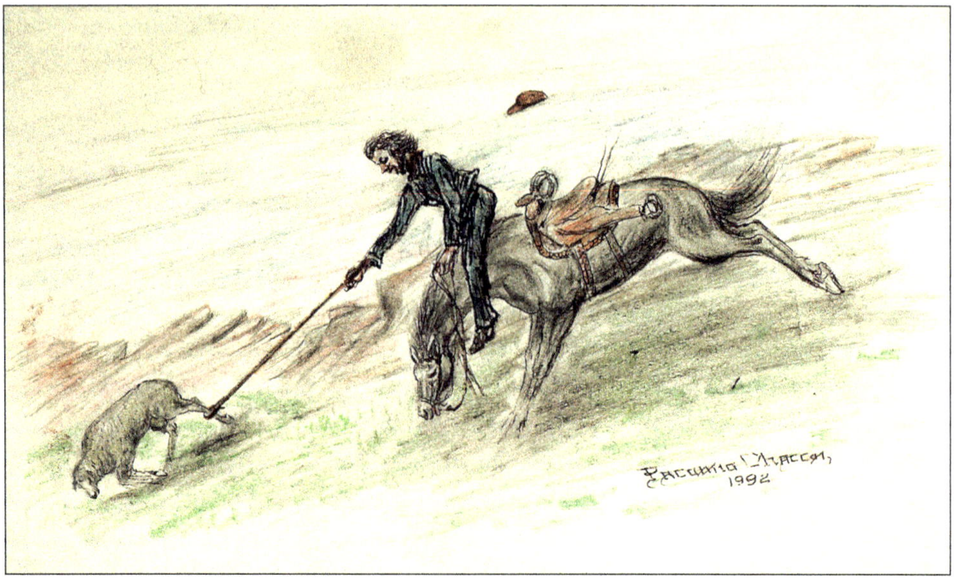

PMAP 1: Paco Chacon's "Unseated Sheepherder." 1992 colored pencil drawing (author's collection courtesy of Paco Chacon).

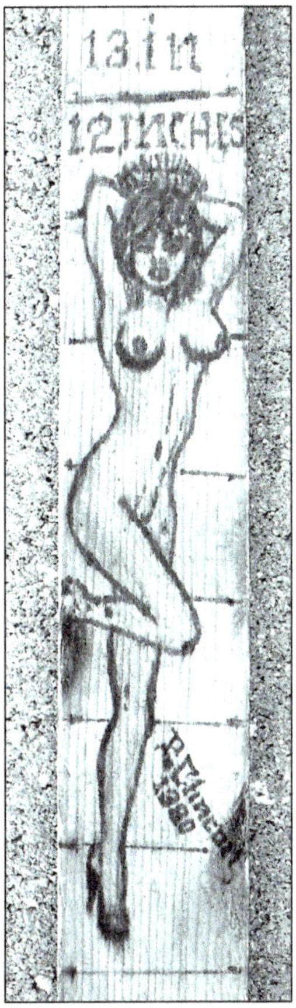

PMAP 2: Details from Paco Chacon's 1980 decorated yardstick drawn with a ballpoint pen for his former employer, Mike Theos (author's collection courtesy of Mike Theos, Denver, Colorado).

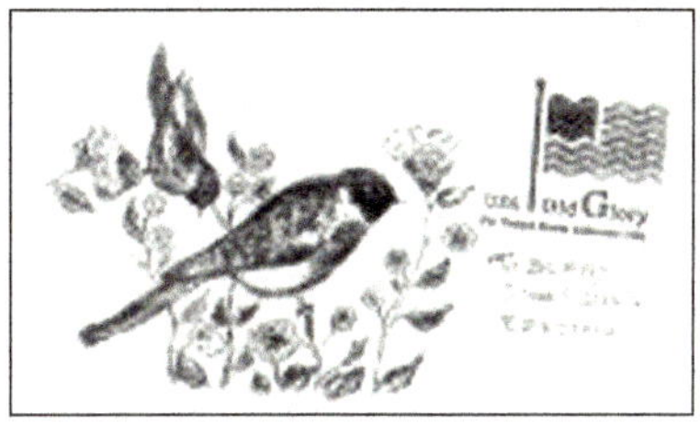

PMAP 3: Paco Chacon's birds drawn with colored pencil for a granddaughter on a mail envelope (author's collection courtesy of Paco Chacon).

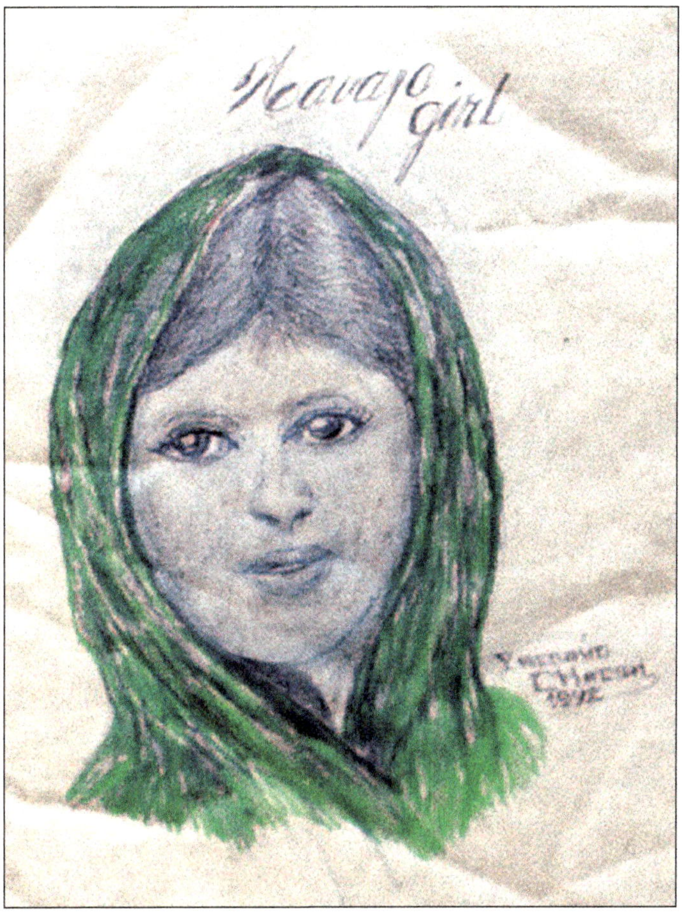

PMAP 4: Paco Chacon's 1992 portrait of a "Navajo Girl." Colored pencil drawing on a padded postal mailer (author's collection, courtesy of Paco Chacon).

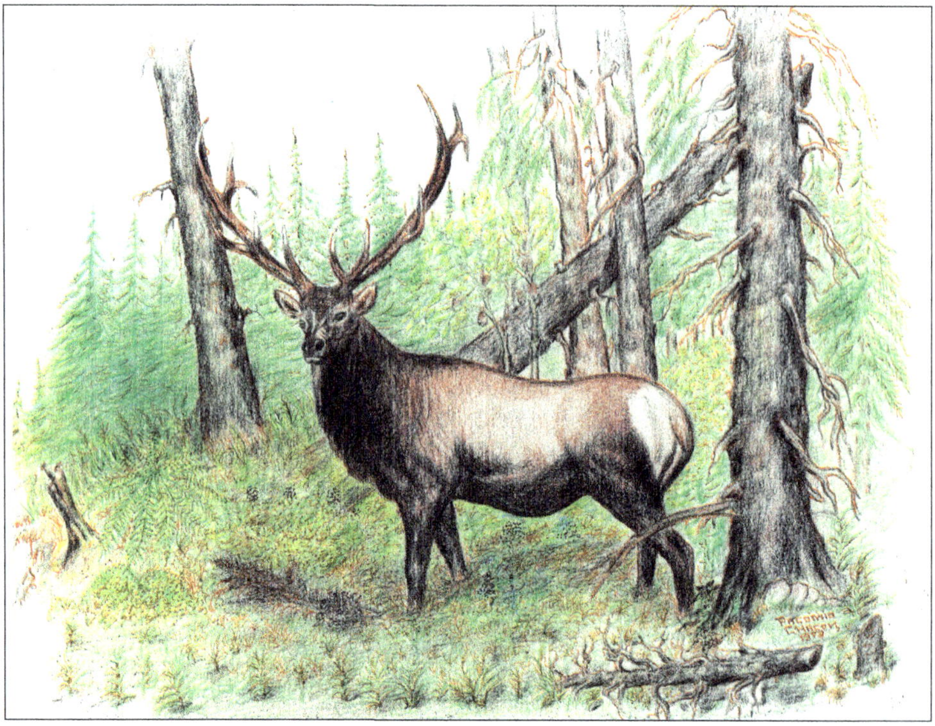

PMAP 5: Paco's 1972 colored pencil drawing of a bull elk in deep spruce timber (author's collection, courtesy of Alice Montano, Grand Junction, Colorado).

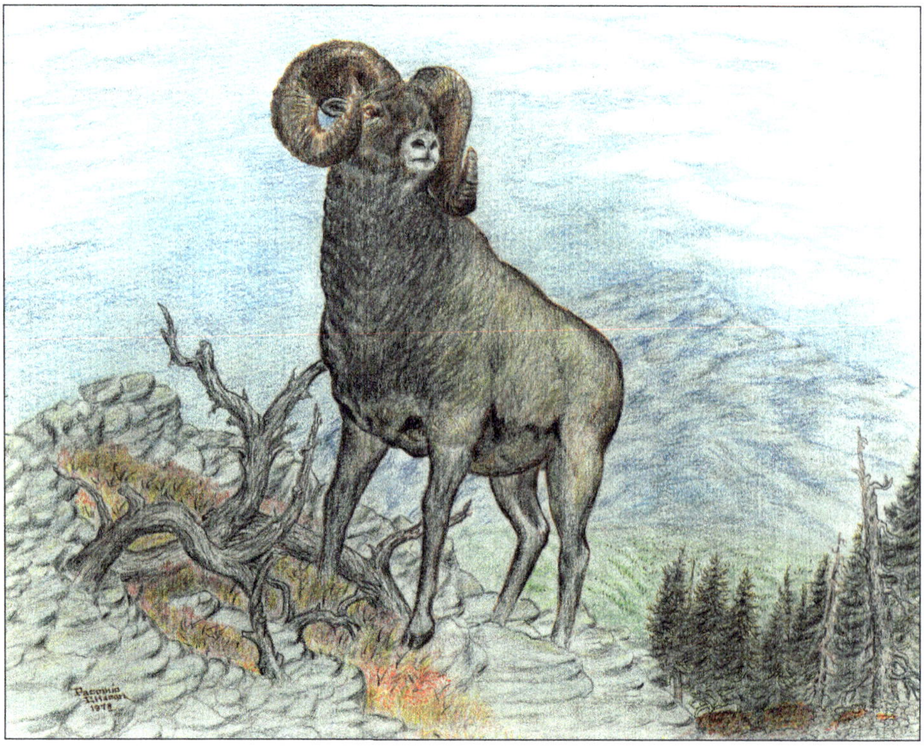

PMAP 6: Paco's 1972 colored pencil drawing of a big horn mountain sheep (author's collection, courtesy of Alice Montano, Grand Junction, Colorado).

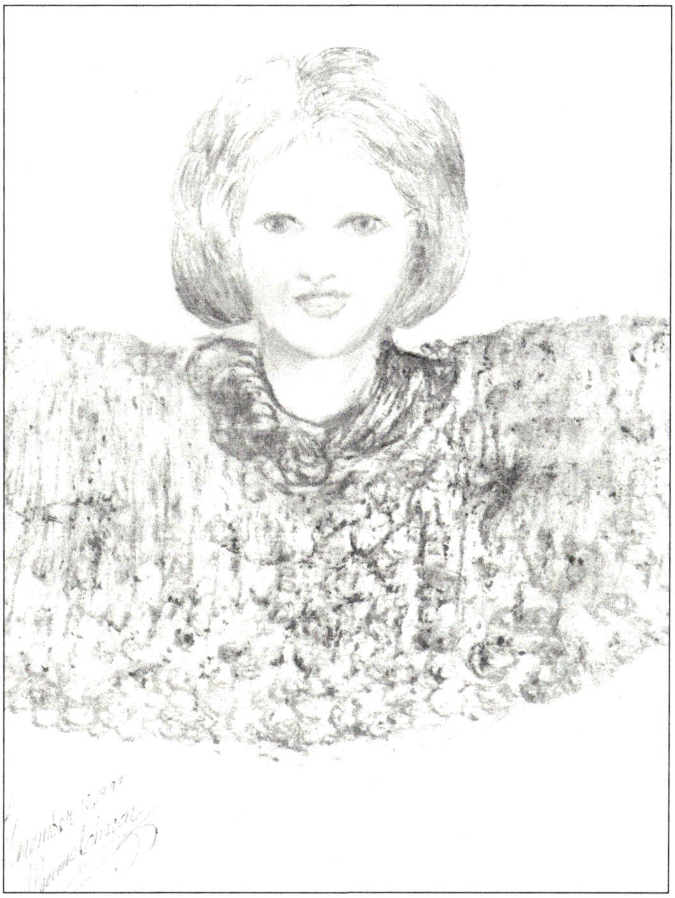

PMAP 7: Paco's 1999 colored pencil portrait of an unidentified woman who may have been a family member. Note the beautiful cursive handwriting on this example (author's collection courtesy of Paco Chacon).

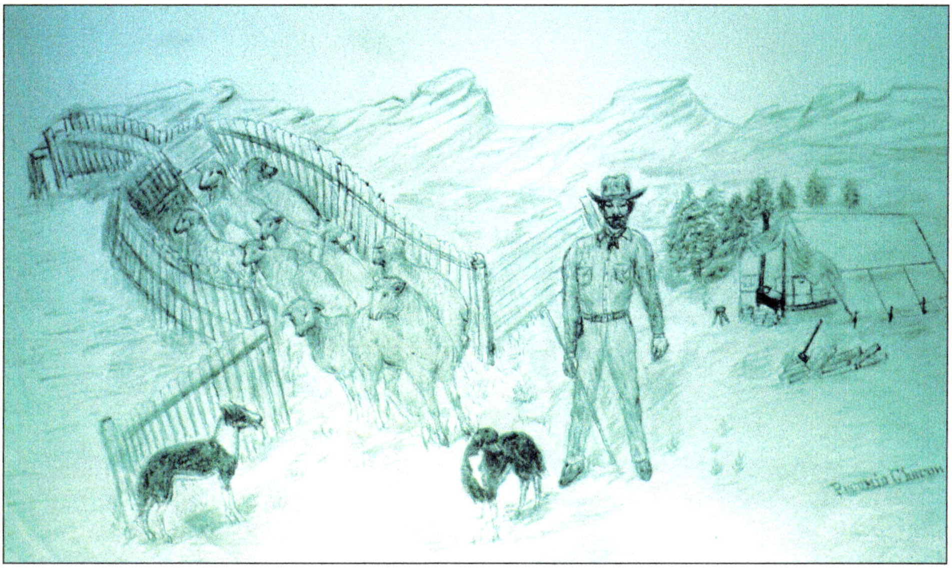

PMAP 8: Paco's undated color pencil drawing of a sheepherder, his camp, and his sheep and dogs (author's collection courtesy of Paco Chacon).

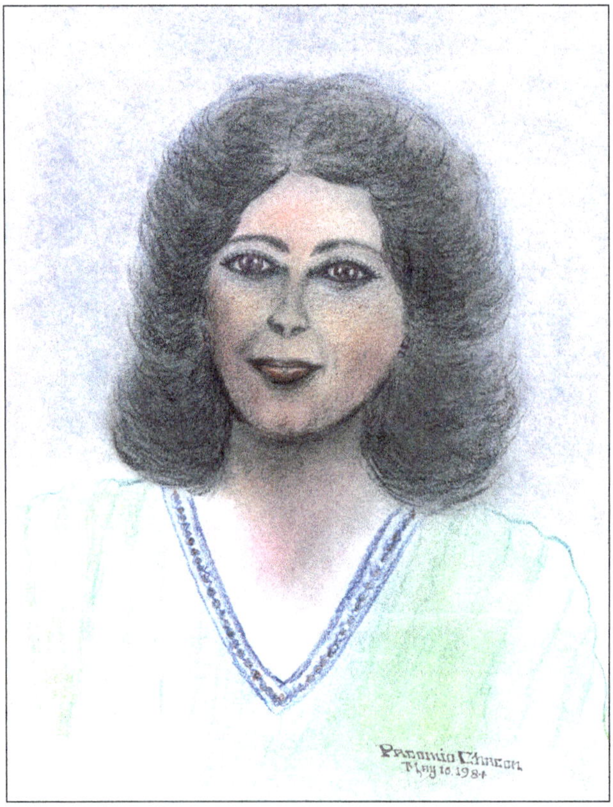

PMAP 9: Paco's 1984 colored pencil portrait of his eldest daughter, Alice Montano (author's collection, courtesy of Alice Montano of Grand Junction, Colorado).

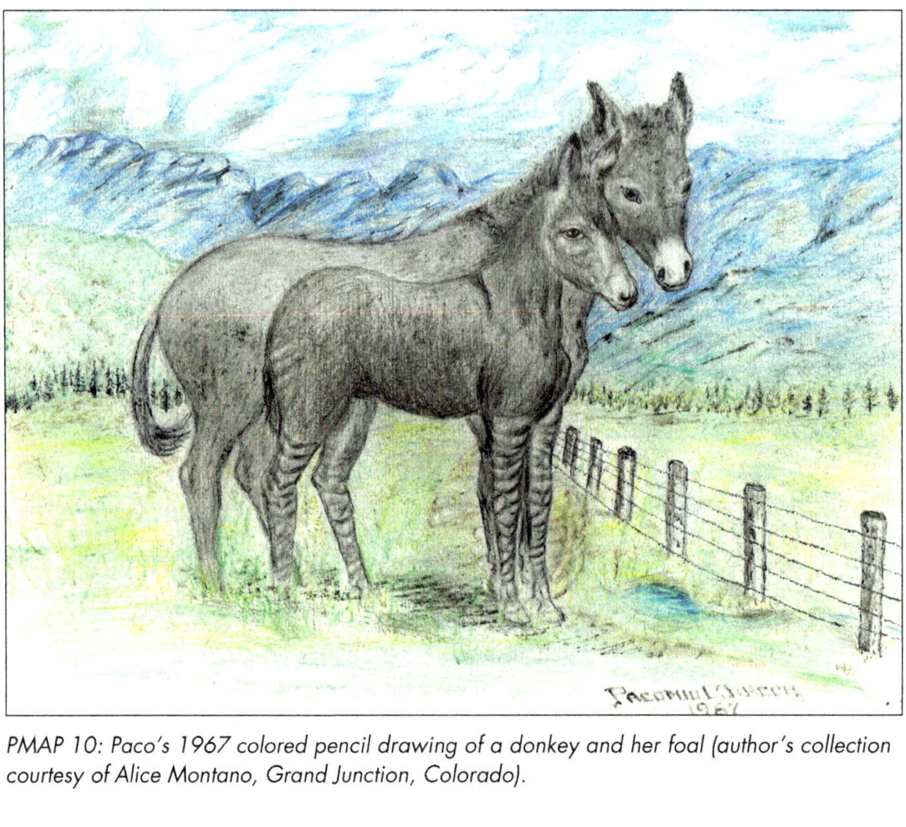

PMAP 10: Paco's 1967 colored pencil drawing of a donkey and her foal (author's collection courtesy of Alice Montano, Grand Junction, Colorado).

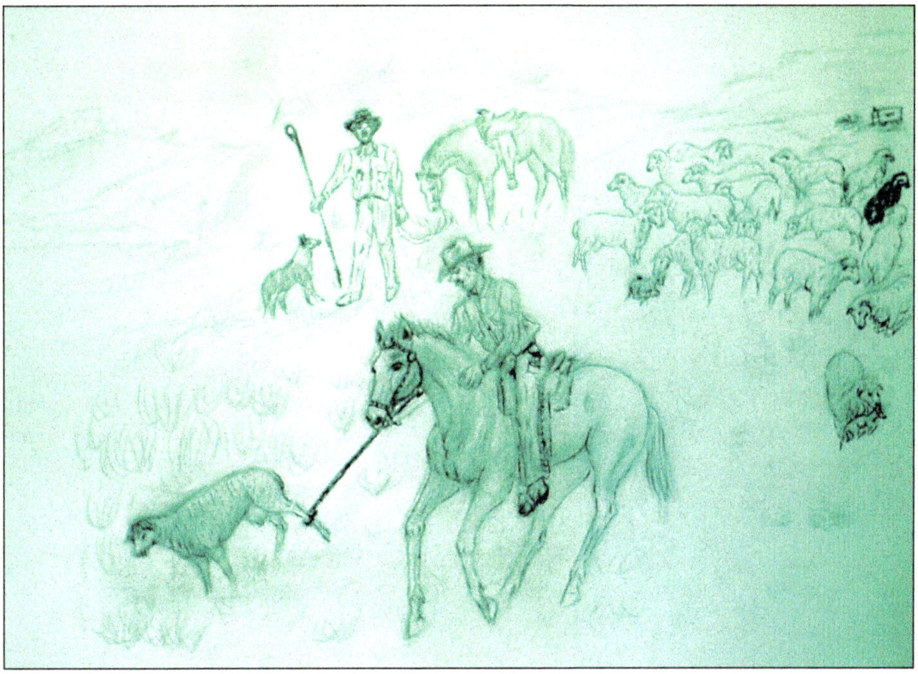

PMAP 11: Paco's colored pencil drawing of sheepherders working with the sheep (author's collection courtesy of Paco Chacon).

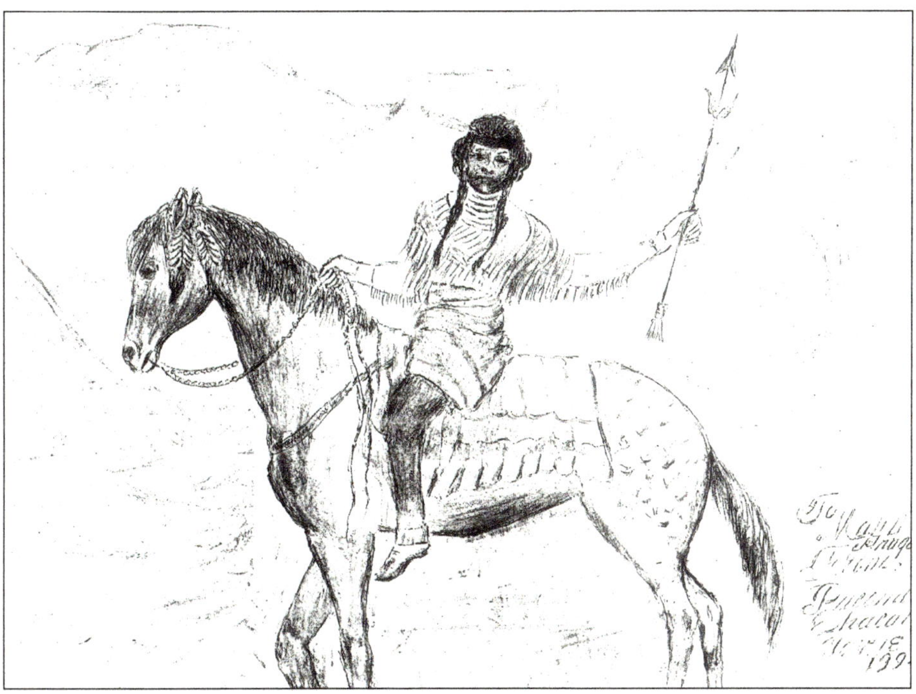

PMAP 12: Paco's pencil drawning of a mounted Native American warrior (author's collection courtesy of Paco Chacon).

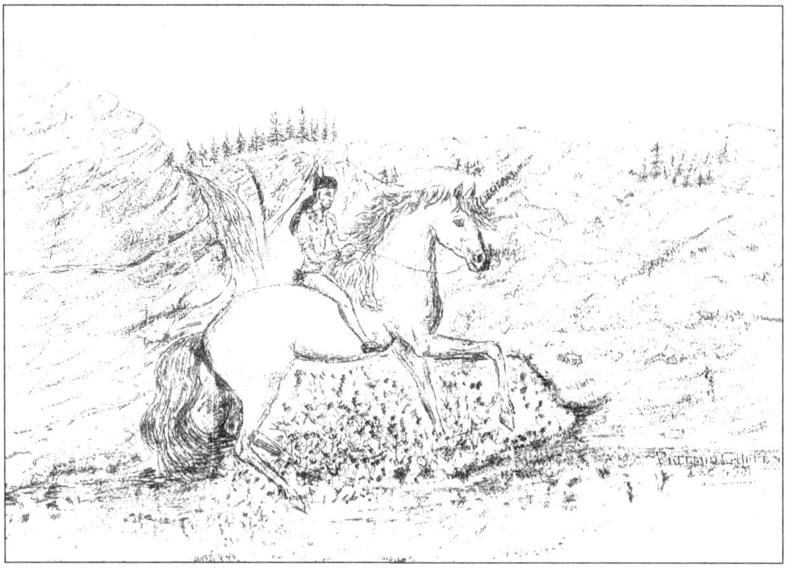

PMAP 13: Paco's pencil drawning of a Native American warrior on a giant unicorn (author's collection courtesy of Paco Chacon).

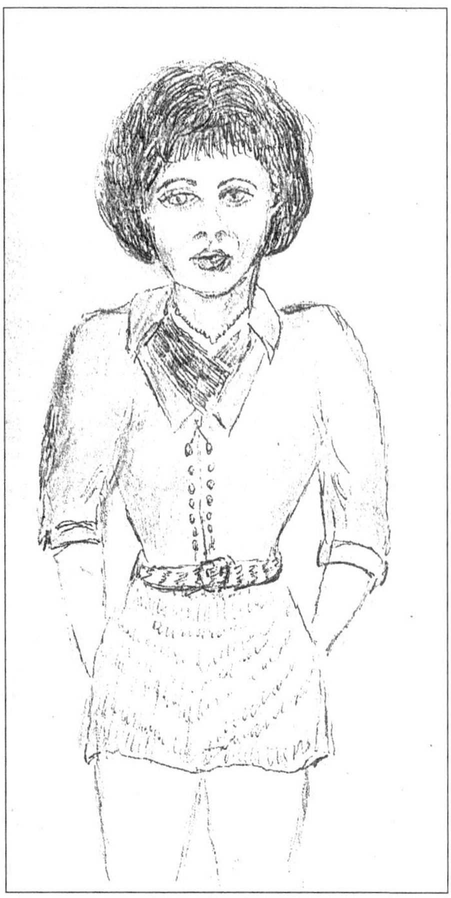

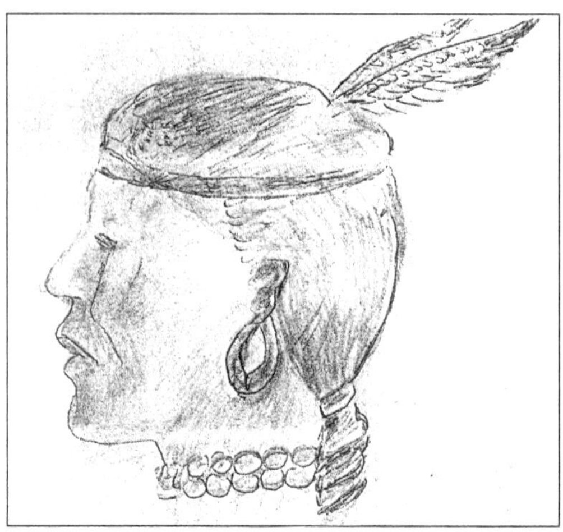

PMAP 14,15: Paco's pencil drawing of an unidentified woman and an Indian man (author's collection courtesy of Paco Chacon).

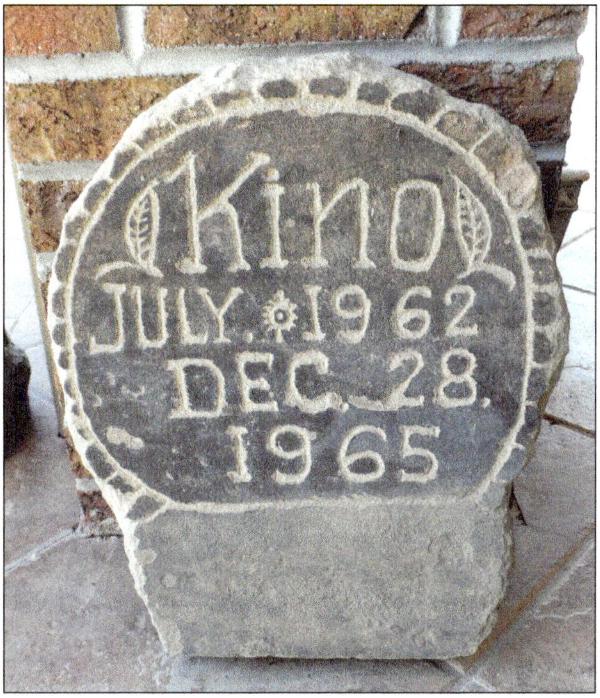

PMAP 16: Paco's carved gravestone for his favorite dog, Kino (author's collection, courtesy of Alice Montano and Priscilla Studt, Grand Junction, Colorado).

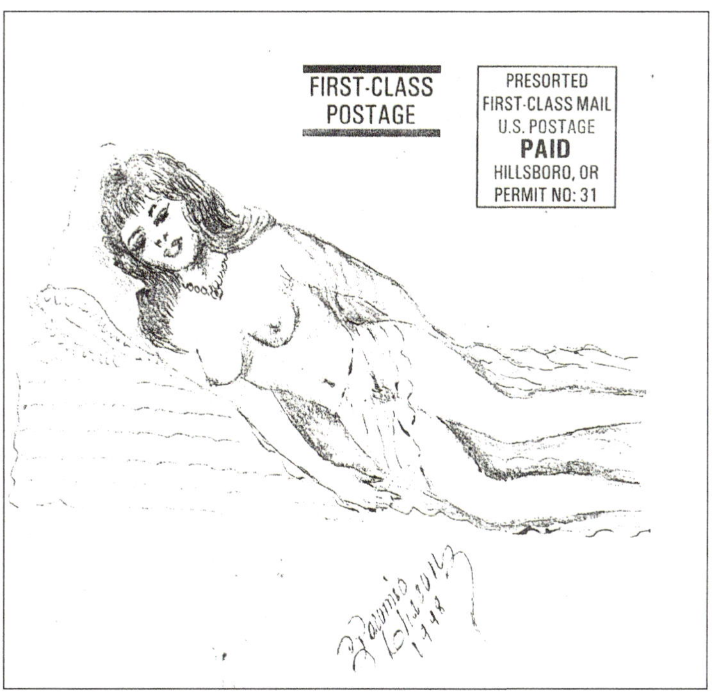

PMAP 17: Paco's pencil sketch of a reclining woman drawn on a postal envelope for the author in 1998 (author's collection courtesy of Paco Chacon).

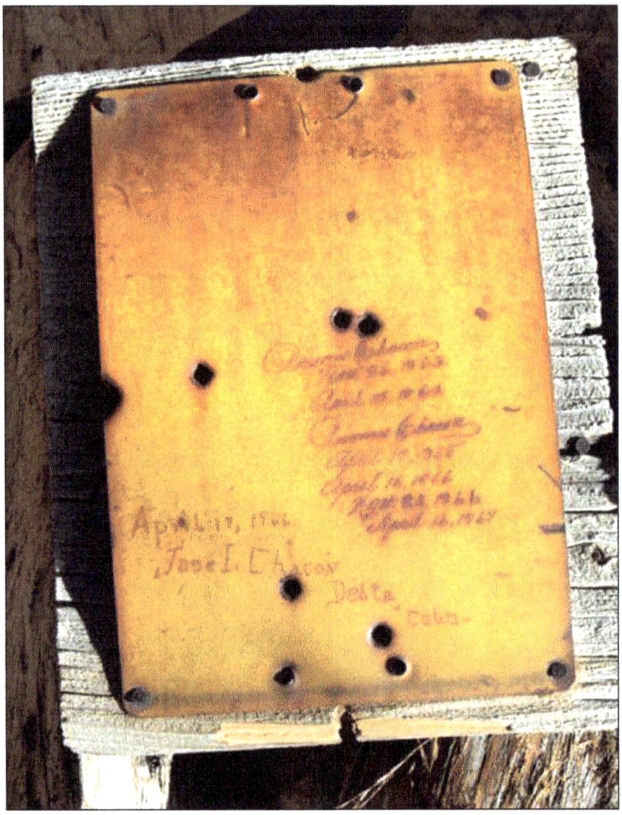

PMAP 18, 19: Detail of inscribed livestock driveway markers from the White River National Forest near Meeker bearing Paco Chacon's signatures and one of his hallmark nudes along with his brother Jose's inscription (author's collection).

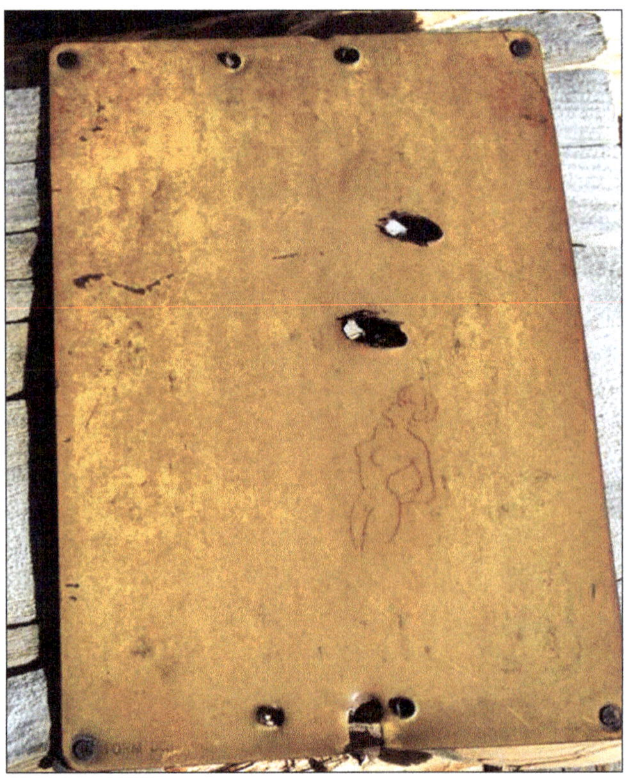

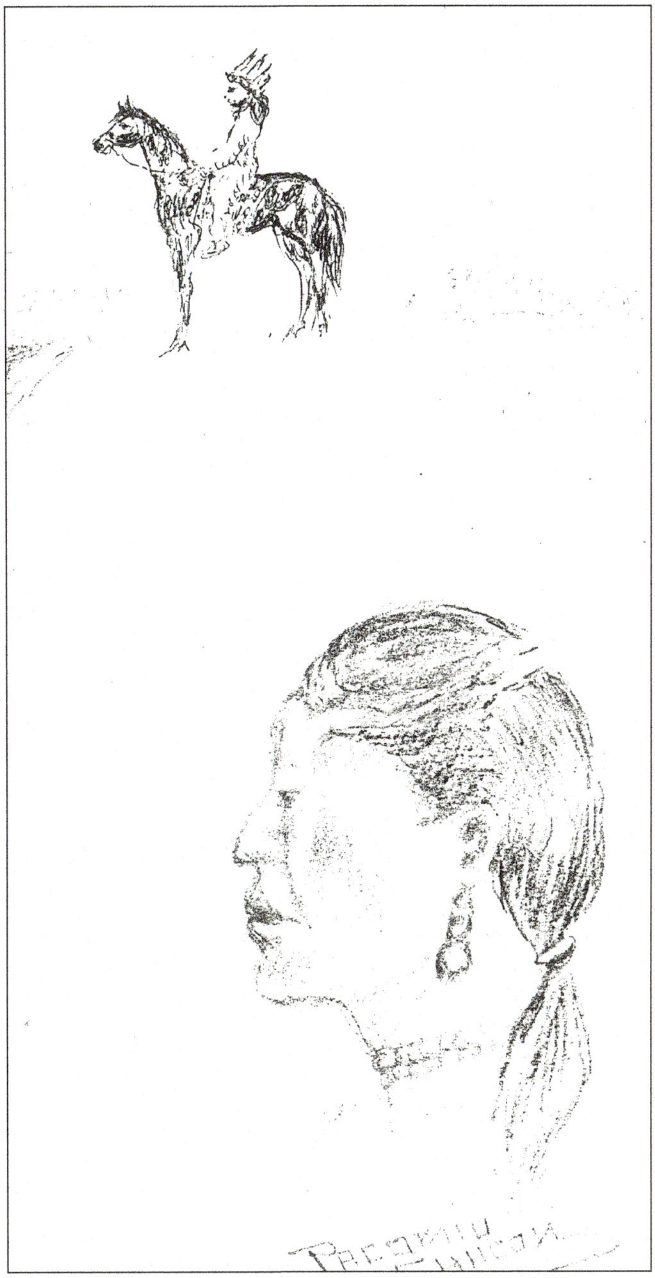

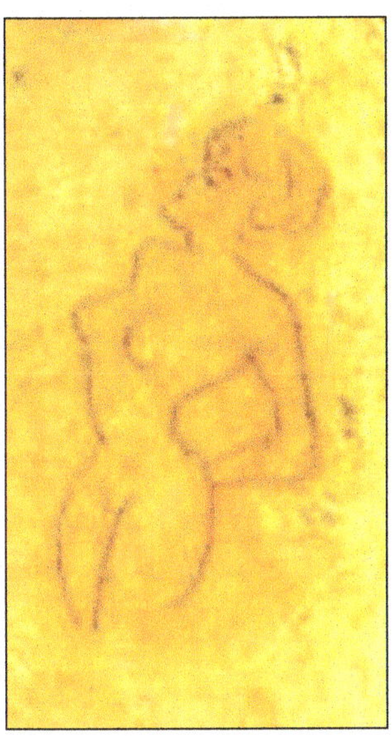

PMAP 20, 21: Paco's pencil sketches of Native Americans and an image of a stock driveway sign as detailed in PMAP 18, 19 (author's collection).

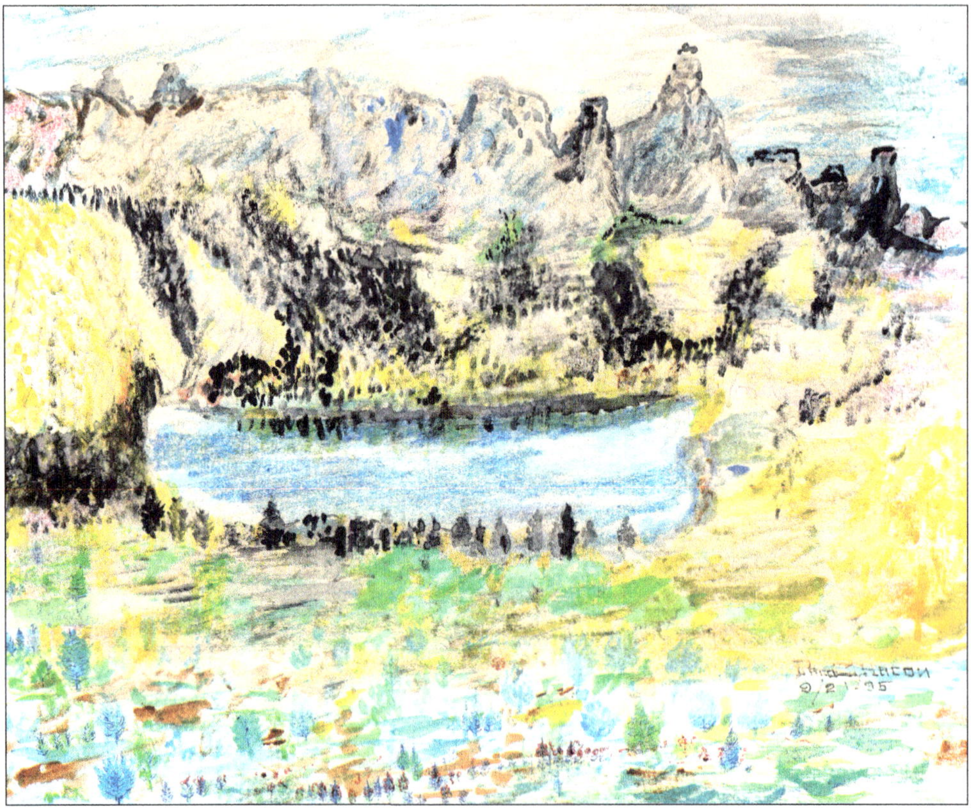

PMAP 22: Paco is known to have made only one attempt to produce a watercolor painting and that was this undated fall mountain scene with a lake. This was most likely painted in the 1970s (author's collection, courtesy of Alice Montano, Grand Junction, Colorado).

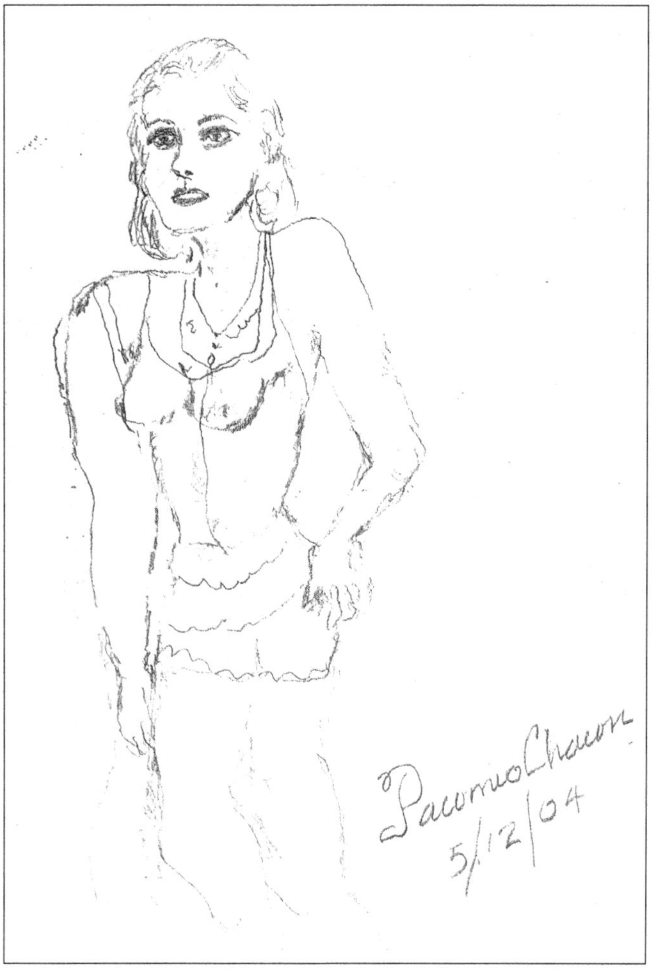

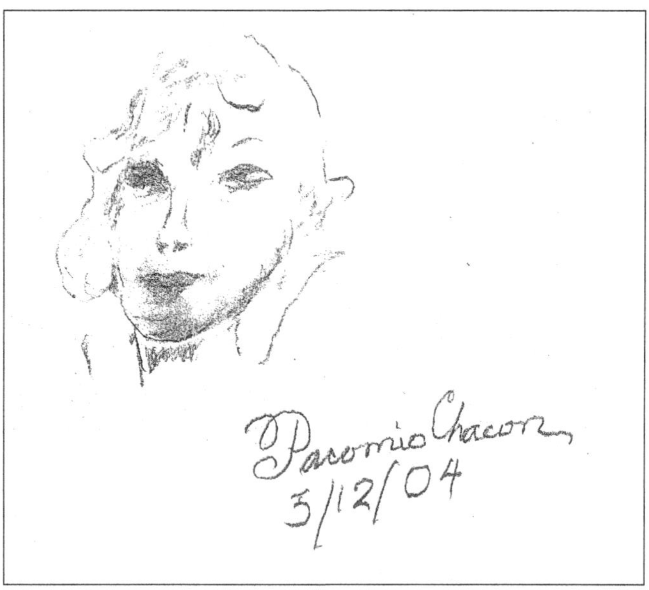

PMAP 23, 24: As he aged, developed Parkinson's Disease, and broke his right hand, Paco still tried to produce pencil drawings for this book. These were two of his last drawings (author's collection courtesy of Paco Chacon).

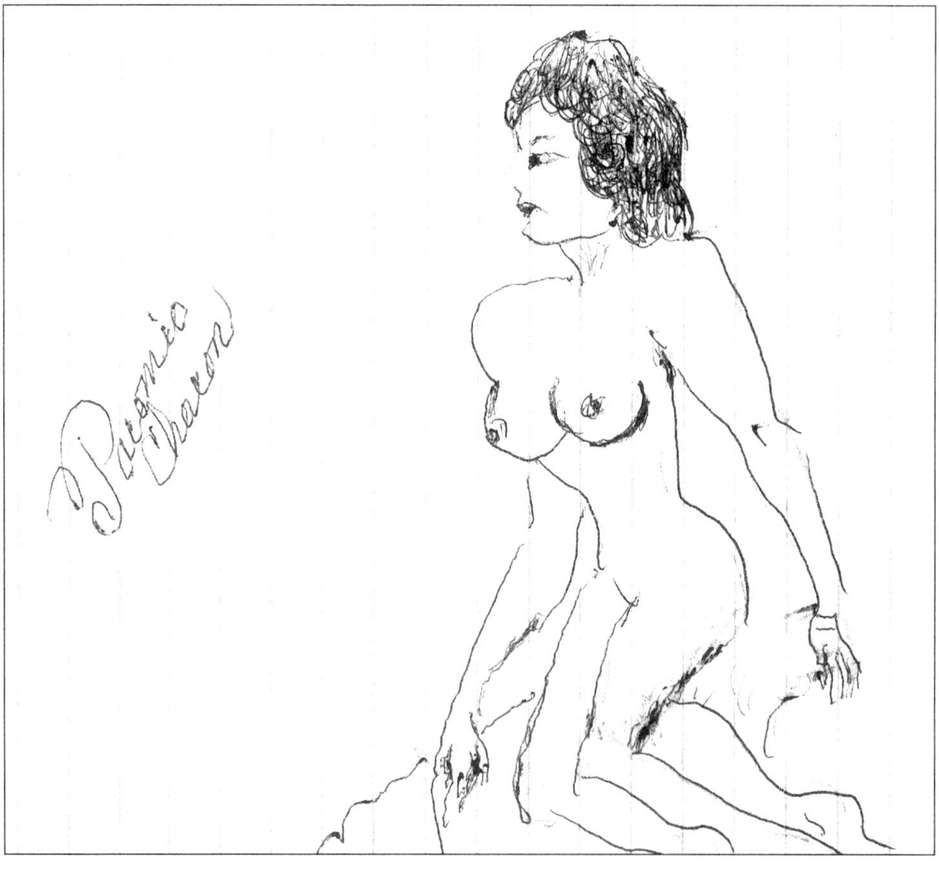

PMAP 25: This ballpoint pen sketch is, to the best of my knowledge, the very last, or one of the last, drawings attempted by Paco Chacon in his effort to assist me in preparing his libro. It was completed about 2006 and closes the record of his run as a folk artist (author's collection courtesy of Paco Chacon).

Chapter 10

EPILOGUE—AN UNLIKELY DUO, PACO AND STEVE—THE AUTHOR'S PERSONAL REFLECTIONS

When I think back on my twenty-plus year association with Paco Chacon, which is often, I am struck by the great contrasts between us. Paco and I easily bridged truly monumental differences in age, cultural backgrounds, and personalities and became good friends. If my readers have read this book to this point they should have by now learned about Paco and the world he hailed from and lived in for his entire life. For the reader to understand our unusual relationship more fully, I believe I must briefly introduce myself.

I am a native of rural Kansas, the great state of the mythical Jayhawk, the mascot of K.U., my alma mater. I am of white, predominantly Germanic extraction, and a descendant of old and poor Kansas homesteading stock. Although I am from a small farming community I was raised as a "town kid" rather than a "farm kid." I am a native English speaker and never heard a word of Spanish until I slept through my Spanish classes in high school in the little community of Holton, up in northeastern Kansas. I had never heard a foreign language spoken fluently until I was in college during the 1960s and traveled to Europe on the then-requisite student backpacking "rite-of-passage" trip to the continent.

I do not know just when one qualifies to legitimately be called an "egghead" but suspect that I at least come close to meeting the bar in that department, particularly in the eyes of Paco and other Hispanic sheepherders. I was professionally trained as an anthropologist and historian and spent an awfully lot of years in universities. I do believe that I ultimately came to be fairly culturally aware and appreciative of other people and lifeways, even though my early years were pretty barren in that regard. I certainly had no experience anywhere along the line with Spanish-speaking sheepherders from the mountains of New Mexico! While I speak a bit of "pidgin-Spanish," on the rare occasions when I ventured to speak a little of it, Paco often slapped his forehead and asked me: "patrón, what you tryin' to say!" I would then abruptly switch to English.

I first became acquainted with examples of art left by some Chacon tree writers in 1976 soon after moving to Montrose, Colorado. That year I was on an archaeological field crew recording some aspen carvings on Lone Cone Peak in San Miguel County for the Colorado Historical Society. At that time I encountered some images, and although they were unsigned, I was impressed by their unusual quality. The names or initials of four or five Chacons were also present in this same grove of trees. These included a P.M. Chacon, which was probably Paco's or that of his brother, Press.

The signatures and images stood out among all the other aspen carvings in that particular grove of trees where there were many other names and images carved. Most were simply the same old run-of-the-mill primitive stuff. I was, however, sufficiently impressed enough to make some duplicates of some of the slides I had to turn over to the Historical Society. Some of these are reproduced in Chapter 6. Although my staff and I completed a great many projects around the region since that time, we never again got into aspen art in Paco's old ranges. I forgot all about it for decades.

In 1989 I initiated a substantial cultural resource study program on the Douglas Creek Arch just south of Rangely in Rio Blanco County on behalf of a gas and oil company from Denver (Baker 1991). At the start I was totally unaware that this work would bring me face-to-face with a substantial volume of Paco's work and into a close relationship with him. During an early stage of this work, my wife, Nancy Ellen, and I were completing a study of a proposed pipeline and road corridor. It was in the middle of summer, the heat was at least 110 degrees, and the black gnats were out. It was simply miserable; Nancy and I were both cranky, to say the least!

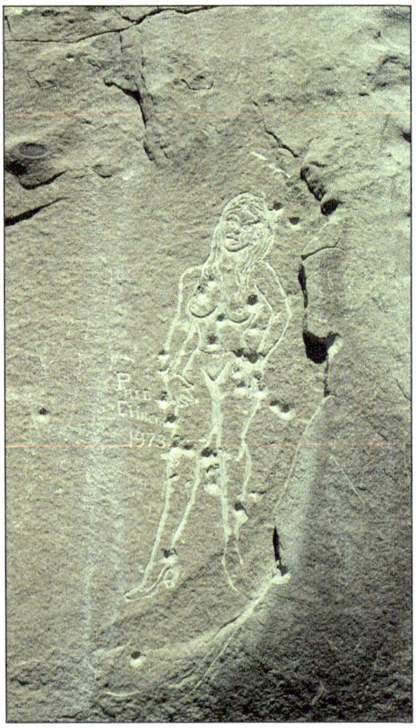

Figure 47: This badly vandalized image, "Paco's Murdered Lady," (PRA 17 in Paco's portfolio) was the first of Pacomio Chacon's petroglyphs that my wife and I found and recorded near Rangely, Colorado in 1989 (author's collection).

As we rounded a corner of a big rock outcropping, I looked up, and there was a sizeable (about one meter tall) and nicely executed carving of a full frontal nude female on a flat place in the cliff face (Figure 47). Although she had once been one of Paco's real beauties, she had been badly shot up by some thoughtless hunter who had used her for target practice. Beside her image was the notation: "Paco Chacon 1973." It only took me a moment to realize that I knew this name from somewhere.

I hailed Nancy, "Hey, come here I think I know this guy!" To say the least Nancy was definitely not impressed that I was all excited about a naked woman carved on a rock! When I asked her to hold the photo scale up against the cliff so I could take a photo, she balked and said she was "not about to have my picture taken with a naked woman!" She was quite adamant and a bit snappy about the entire matter. I do not know just how we worked it out but--perhaps by mustering all my charms--I ultimately persuaded her to hold the photo scale without having to resort to overtly reminding her that I was still the project archaeologist and thus still the boss, at least until quitting time.

A bit later I was discussing this carving with the archaeologist who was supervising my work on behalf of the Bureau of Land Management (BLM) that had issued the permit that I was then working under. He told me that I should "know how to record these kinds of images and to just take care of it." At that time the Forest Service was not yet paying much attention to this kind of carving among the aspens. The BLM was seemingly not sure of just what to do about such carvings on the cliff faces since they were less than fifty years old and thus not at all eligible for listing on the National Register of Historic Places. Although Native American rock art is commonly encountered and recorded on the agency's lands, sheepherder rock art, and particularly images of naked women, only a few decades old was certainly not typical of the kinds of cultural resources that archaeologists routinely deal with then or now.

As our work progressed in Southwest Rangely Field, my study team began to record more and more of Paco's cliff carvings, and the BLM was seemingly both approving and appreciative of our efforts. My crew and I were impressed. This was obviously good folk art. One of my key crew members, Phil Born, had worked as an archaeologist for the Forest Service and had recorded a number of Paco's aspen carvings. The Forest Service had begun to take notice and had already sent Phil and his supervisor, Polly Hammer, to interview Paco. By the 1970s word was starting to get out about both Paco's cliff and aspen carvings within the archaeological circle. It was rapidly becoming obvious to all that Paco's work was truly exceptional and that he was an important person among the sheepherding cadre of tree and cliff writers.

Part of my study program in Southwest Rangely Field involved recording a wide variety of old sheepherders' camps and other features. The carvings Paco and a few others left fit nicely into this work and promised to contribute to its ultimate success as an archaeological study. This led me to make a point of meeting Paco and getting to know him. He was by then pretty much retired so I occasionally would stop in Fruita and visit with him at his home. While we were becoming acquainted, other people were becoming more aware of his art.

Although I can no longer recall just how it came about, I was put in touch with Gus Halandras of Meeker, Colorado. Gus descends from old Greek sheepherder stock in that region and was instrumental in developing the famous Meeker Sheep Dog Championship Trials, which are now held annually. Like most local sheepmen, he was well aware of Paco's art. An invitation was thus extended for me to bring Paco to the trails in 1992 so that he could be formally acknowledged by his peers in the sheep industry and give demonstrations on aspen carvings.

We subsequently received a grant from the Colorado Council for the Arts. This was obtained on our behalf by the Museum of Western Colorado through the good auspices of its director, Mike Perry, and Ronna Lee Sharpe, the staff folklorist. That grant provided funding for me to take Paco into Southwest Rangely Field, visit his carvings, and interview him. It also provided funding for me to develop a professional-quality art show of his work and take him and the show to the trails in Meeker in 1992.

In the spring of 1992 Paco (Figure 48) and I spent some days knocking around northwestern Colorado and revisiting much of his rock art. We were getting to know each other and began talking about all kinds of things from sheepherding to our own family lives. We found it easy to talk, got along well, and began to share a lot of personal confidences. In September we left for Meeker. It was then that I got my first big lesson in dealing with Paco and his brother, Press.

With no intention of sounding harsh or disrespectful, suffice to say that if there was a Chacon who easily fit into the common perception of the "crazy sheepherder," Press easily fit that bill, God rest his soul. Interesting, entertaining, nearly always personable, knowledgeable, and wholly irascible, he was quite a character. When he and Paco began drinking, you could readily find yourself with, to say the least, "quite a handful!"

After arriving in Meeker, Paco suggested we go out to the Yellow Jacket summer range and see Press who was herding sheep there. Sounded like a fine idea to me so out we went, arriving late in the afternoon. We all sat down in the sheep camp trailer, and Press offered diner, which was a lamb stew as I recall. After dinner Press produced a substantial bottle of whiskey and offered us all a drink. My view, was "don't mind if I do" since I am not known to ever be averse to a good dram at the end of the day.

After a few rounds, it was time for me to go out to my camper and go to bed. Paco and Press were still tipping her up when I turned in. Unbeknownst to me, the two just kept at it until there was not a drop left anywhere in camp. About 3:00 AM I had to get up to water the bushes. I got outside just in time to see Paco silhouetted in the door of the camp trailer. He then pitched forward and did a near perfect "header" (I cannot call a swan dive) out of the doorway. He hit the ground with an audible thud.

After checking him to see if he was alive or had any broken bones, I thought, "Oh Hell, now we are in it!" We had taken money from the Arts Council and promised to appear at the dog trials by 9:00 AM and the star attraction was lying in a mountain meadow pickled to the damn gills! I finally managed to get Paco up onto the floor of the trailer and covered him up with an old blanket. In the morning he was up but moaning, "Oh I think I am going to die, ay ya ya!" He said we should go down to his old boss's place, Mike Theos' ranch, and get some coffee and something to eat.

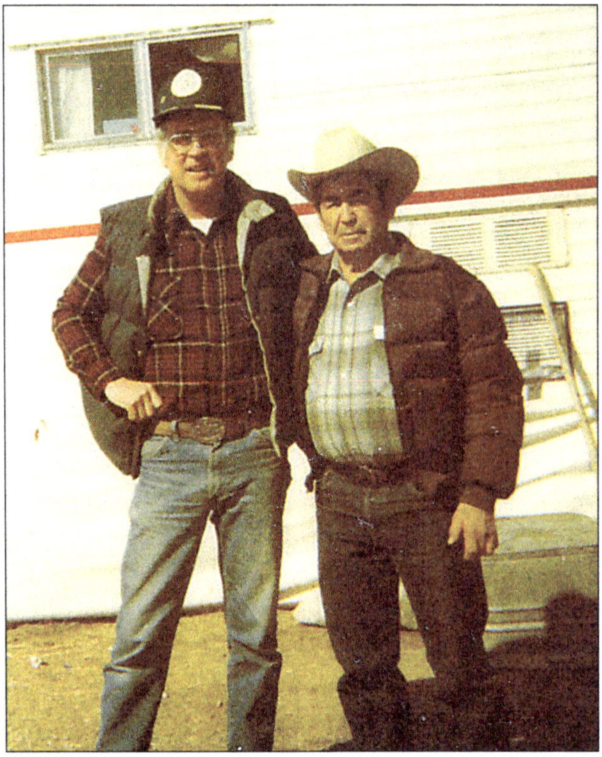

Figure 48: Steve Baker and Paco Chacon in the spring of 1992 at the winter sheep camp of Press Chacon south of Bonanza, Utah (author's collection).

We then drove to Mike's house, and he filled Paco with coffee and toast. All the way into Meeker, Paco was afraid he was going to be sick. All I could think of was how silly I had been to take him to see Press. But then I was not yet fully aware of just how much the two men could and would drink whenever booze of any kind was available. I was also concerned that Paco might vomit in my pickup.

We got to the trials a tad bit late, and all Paco could do was sit in the truck and sleep. I made some excuses for him and before noon finally had him out and entertaining the visitors (Figure 39). This was a big lesson for me. Paco and I subsequently, on my insistence, agreed that he would not drink when we traveled together. It became an iron-clad rule.

Late in 1993 my research on the Douglas Creek Arch ended when my old and most-favored client of fifteen years abruptly changed management. The new administration summarily fired all of its employees and consultants who had thoughtfully labored to bring it profits for many years, including me and my company, Centuries Research, Inc. Several, including Centuries, were even sued under the most specious of circumstances. In my case the new administration did not want to pay the costs for me to produce the hefty archaeological reports then still owed to the BLM. Incredibly, it also very foolishly believed that it owned all the records and artifacts I had recovered from federal lands over the previous years under Centuries' antiquities permits.

Along with accusing me of absconding with the company's artifacts, one of the salient points the lawyers seized on to claim I had unwisely spent their client's money was the fact that I had been recording Paco's carvings of naked women! The BLM, of course, did precisely as one would expect and executed a perfectly choreographed bureaucratic ballet. It performed a classic "stall maneuver" and literally whirled in backward circles. Although it had always approved of and formally permitted our work, the agency ultimately hung me and Centuries out to take the rap alone. It would neither affirm nor deny that their representatives had instructed us to, among other actions, record Paco's work.

Although the wholly meritless but expensive litigation was eventually abandoned by my former client, it had involved Paco's artwork in a substantive way. I distinctly recall the day that I and my former great boss, Hugo Cartaya, from my client's firm (who was also fired and sued) gave our depositions in a fancy law office in downtown Denver. At that time an obviously blood thirsty oil company attorney, literally oozing with malice toward archaeologists, waved a copy of a picture I had taken of one of Paco's ladies in front of my eyes. He gruffly asked me if my client had paid me to take photos of "this kind of stuff?" My response was strongly affirmative. I knew I had been conducting myself professionally and doing the right thing.

Although my Rangely program had ended on a sad note, I continued to see my friend Paco whenever I was in Fruita and on one occasion brought him home to Montrose to spend a day, meet my wife and young boys, and have dinner. Paco instantly took to Nancy, Christian, and Matthew and enthusiastically talked with them about sheepherding. Nancy and I have a nice grove of smooth skinned young aspens in our front yard, and Paco was admiring of them.

While he was looking the aspens over and telling us what good canvases they would make, my comely wife was becoming increasingly nervous for fear that Paco might attempt to replicate her likeness on one of the trees. While I knew Nancy would not model for him, I figured Paco would have a pretty good idea about just how to size her up in her most beautiful natural form and render a good likeness! I thought it would be a pretty dandy idea and would have approved in an instant! How many men can boast that they have a likeness of their beautiful wife carved by none other than Pacomio Chacon on their own tree? I was going to bring Paco home again just so he could do some carving, but it never happened. I still hope to make templates of some of his best works and use his techniques to replicate them on my trees.

Paco had always called me "patrón" after our first meeting. I tried to dissuade him from doing so, but he insisted and always addressed me and introduced me to his friends as "mi patrón." I told him that if he insisted on calling me patrón I would call him "el profesor de las ovejas" or the professor of the sheep. Thus it became "Patrón, Profesor, Patrón, Profesor..." all along our routes. By calling me patrón, Paco was showing his respect for me as his sponsor and spokesman and not as his boss under the common definition of the term.

As we grew closer and Paco came to realize that I was truly going to write a book about him, he began to take on an exuberant shine and smile whenever the subject came up. He proudly told his friends that his patrón was writing a book about him. In 2004 Paco wanted to see his old home at La Mesa one last time. He had not been back there for many decades. That year I took Paco to La Mesa to meet some members of his extended family and see where he grew up.

One evening heading to La Mesa we were in a small motel in Alamosa, Colorado. He asked me: "hey patrón, can I have baaath?" Paco loved to take baths whenever we traveled. I responded, "sure Paco, have a bath." Then I asked him if he needed help. He said he did. I got him safely into the tub as he was then in his early eighties and growing increasingly infirm. He then asked "hey patrón, can you wash my hair for me?" I felt honored that he would ask me to help him in this personal way.

I said sure and began to shampoo him up. He was just a tiny, shriveled up old man sitting in a tub while I poured water over his head. It was then that he looked up at me with his tired old eyes and said, "hey patrón, are you really gonna write a libro about me?" I responded "yes Paco, but it is going take some time. I still have to gather all of your pictures from the woods and canyons. So you can't die until we get it done." Paco's response was "ok patron, I won' die." He then asked me, "hey patrón, can I have a beer with my supper?" Although I was certainly ready for a beer myself, I had to remind him that we did not drink when we traveled. That particular situation and exchange with my good friend is indelibly imprinted in my mind and will be with me all the rest of the days of my life.

It is a one-hundred-and-fifty-mile round trip to Fruita from my home in Montrose. When I was in the area, which was not often after the end of the work in Rangely, I stopped and saw Paco and Press. Perhaps I had dinner with them and tried to ignore Press's constant requests for me to take him to the liquor store. Neither he nor Paco ever drove and could be very pushy when they needed to go buy booze. I acquiesced to their requests once. That experience with Press was a nightmare. I was truly afraid I was going to be arrested and named as an accomplice to his overt sexual harassment of a certain female liquor store clerk in Fruita! Press simply would not keep his hands off that overly patient woman who, thankfully, did not call the cops. Wisely, I never made that mistake again!

Whenever I saw Paco, he asked if I was still going to write the book about him and his "pictures." The answer was always the same, "yes," but only after I find the pictures. "So Paco, remember you can't die. You have to live to see your book." His response was always "ok patrón, I won' die." This little ritual played out over a number of years, and it became emotionally harder and harder for me to keep it up. I had a family to support and was often simply overloaded with paying archaeological research and writing that had to be done.

I was becoming increasingly aware that I might not get the book written while Paco was alive. Paco's pictures were all scattered out in the woods and rocks over thousands of square miles, and Paco and I were both getting older. Paco still drew some on paper or cardboard and tried to save the images for me. They just kept disappearing or he would give them away to his friends when drinking.

Although I have lost track of when we did it, Paco and I made one last trip together. I think it was about 2006 soon after he entered the assisted living facility at The Oaks in Fruita. Paco and Press had often spoken of the wonderful carvings he had left in Tabyago Canyon just off the Green River in eastern Utah. I was wholly unfamiliar with that region, and Paco thought he could show me how to access the canyon even if we might not be able to walk in and find the carvings.

We set out in the pickup for Utah and drove and drove and drove. Paco was unable to find his way since evening was coming on, and there had been a lot of gas and oil development in that area since he last worked there. There were many new roads, and the land had been folded into the Northern Ute Indian Reservation. The day grew long, and Paco was very tired so we went to a motel in Vernal. Paco did not eat much at dinner and could not sleep. He spent much of the night pacing about and wanting to call his daughter, Alice. The next morning we went home and agreed that we would not try to make any more trips. Paco was just too old and frail.

After he entered the assisted living facility, he still tried to draw. He developed Parkinson's Disease, and his hands grew shaky. He then fell and broke his right hand, and he was never again able to draw much although he still tried so that I would have something for his libro. One day I stopped to see him; he was ill and vomiting into his waste basket. He said, "patrón, I think I am dying." I reminded him that he could not die until the book was finished, and I went and got a nurse. She gave him some medicine. I was quite concerned about Paco so I stopped in a few hours later to see how he was getting along. Lo and behold! He was sitting up in the dining hall drinking coffee with his buddies. Once again he introduced me as "this is Mr. Baker, mi patrón!"

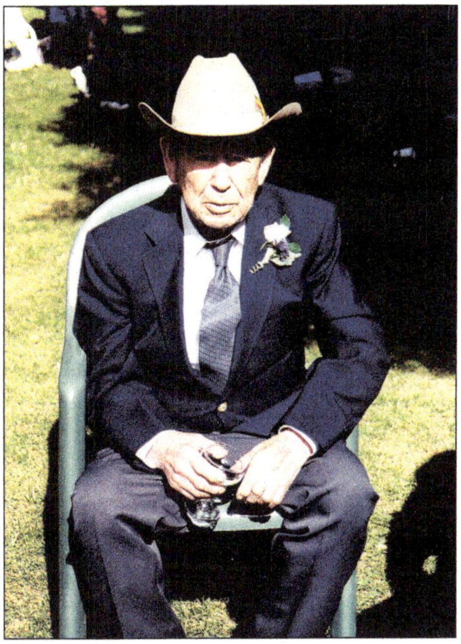

Figure 49: One of the last photos of Pacomio Chacon taken at a family wedding ca. 2006 when Paco was about ninety years old (author's collection courtesy of Alice Montano, Grand Junction, Colorado).

As with so many old folks, Paco's last years (Figures 49, 50) were not particularly good ones. He developed dementia and had to be moved to a secure facility in Fruita. He wandered, and the staff, his family, and his friends put up his pictures and images of sheep along the hall to try to lead him back to his own room. The last time I had a decent visit with him, he was sitting on a padded bench in the hallway. An elderly lady was on the other end of the bench, and she was busy peeing her pants. The yellow river was cascading down a crease in the plastic cushion cover heading straight for Paco. I said "Hey Paco, why don't you stand up before you get all wet!" He responded, "Ok patrón, I don' wan' get wet," and he quickly stood up.

As I left him that day, he walked me to the locked door. Just as I was being buzzed out, he walked over right in front of me. He knew he was near the end of his life. He pulled himself up as straight as his then badly shrunken, barely five-foot frame allowed. He placed his hand in mine, took my other arm by his, looked me square in the eye, gave my hand a good squeeze, and said: "Thank you patrón for coming to see me. Please tell your wife and sons that I remember them and said hello." I left and never again was able to communicate with him meaningfully.

When my own father grew old, he cheated death on several occasions before dying at eighty-nine. I thus lived in a form of denial, always thinking that dad would ultimately escape the Grim Reaper once again. It was much the same with Paco. I wanted so desperately to write his book for all those years. I wanted him to live to see it and take heart in his accomplishments. In my last few visits, he was fully in dementia and did not make any sense. We were never again able to joke about how he could not die. He did though. He never saw his libro and that is my very heartfelt and great personal tragedy. It will haunt me to the end of my days. I strongly believe in honoring one's promises. Although it is a key element in my personal value system, I could not fulfill my promise to Paco! All I can do now is complete his book and stamp it here with a bit of verse to honor my friend's life passage.

Soon I will pass o'er the Great Divide
For I hear Thy welcome hail;
Oh, Master, guide and forgive me all
At the end of the Painted Trail.
(Extracted from the Painted Trail [Fletcher 1986:42])

Although his life force had been extinguished, Paco and I had one last laugh together. His family asked me to assist at his funeral. This was held at the Assembly of God church in Fruita on July 25, 2009. In addition to preparing and reading a eulogy for Paco, I was asked to bring some of his framed artwork from the show we had prepared for display at the Meeker Sheep Dog Trials. As I was setting up his exhibit in the vestibule of the church, the pastor became

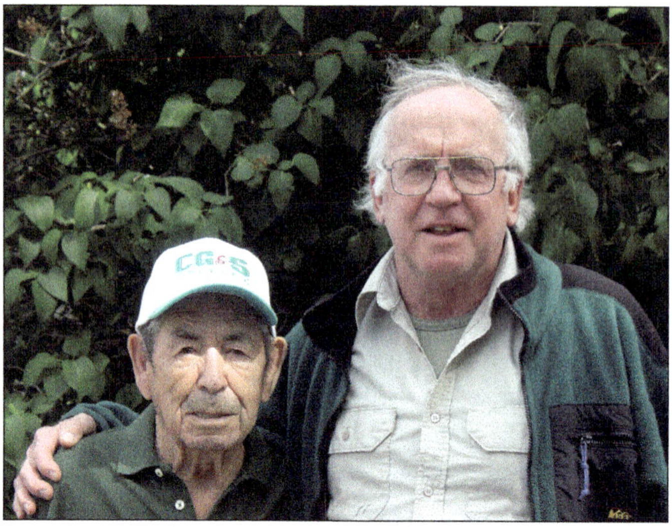

Figure 50: Pacomio Chacon and I at Paco's old home at La Mesa, New Mexico in 2004 (author's collection).

agitated and asked me what I thought I was doing. He told me that I could "not put up such kinds of pictures" in his church.

Little did the good pastor realize that I had just enough good sense and notions of propriety left in me to have high-graded the collection and removed all the questionable items. The one woman in the lot was fully and very tastefully clothed. This is PAP image 2, "Paco's Aspen Lady" from his aspen art portfolio in Chapter 7. I reassured the pastor, and we had a nice exhibit as testament to Paco's artistic skills. Some of the family and I had a pretty good chuckle over this little last event, and I am confident that Paco would also have thought it pretty funny.

Paco's obituary appeared in the *Grand Junction Daily Sentinel* on July 24, 2009 (Daily Sentinel 2009). He was survived by his six children, fifteen grandchildren, twenty great-grandchildren, and eight great- great- grandchildren. Hopefully some of these descendents will manifest whatever family gene it was that gave the artistic ability to Paco. Paco's remains were interred with full military honors at the new Veterans Memorial Cemetery in Grand Junction. I was, fortunately, able to attend. I do believe it would be nice if his clean, white military marker (Figure 51) could carry at least a small image of his artwork. That is, however, probably not likely to happen in a military cemetery unless someone just happens to find an opportunity to place one just below the sod line where it will not be easily noticed.

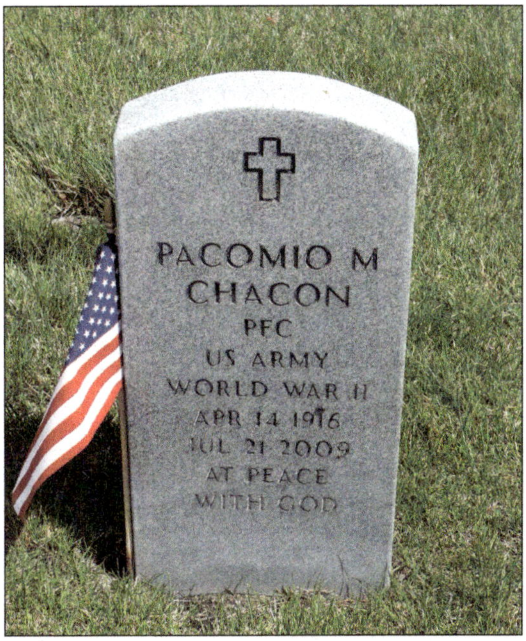

Figure 51: Paco's grave marker in the Veterans Memorial Cemetery in Grand Junction, Colorado. If readers care to stop by they will find Paco in Section 8, Row S, Grave No. 24 to the left immediately inside the front gate of the cemetery. View of May 2013 (author's collection).

Paco's story has now been told. The little sheepherder is dead and buried but his artwork will now live on. I am gratified that I have assured that and my promise to my friend has finally been kept, even if late. Now that it knows what to look for, all that remains to be done is for Paco's newly informed public to get out into the woods and canyons and find more of his art work. There is certainly much that still awaits discovery!

REFERENCES CITED

Associated Press

2000 Fighting the Clock: Forest Service Tries to Document Tree Carvings. *The Grand Junction Daily Sentinel*, March 26, 2000.

2013 Utah Shepherd Gored by Elk. *The Denver Post*, September 8, 2013.

Baker, Steven G.

1991 *A Five Year Research Design and Treatment Plan for Cultural Resources in Chandler And Associates' SW Rangely Area of Undertaking, Rio Blanco County, Colorado.* MS, Chandler S.W. Rangely Field Cultural Resources Report No. 32. Centuries Research, Inc., Montrose, Colorado.

2003a Interview with Pacomio Chacon, Fruita, Colorado, February 5, 2003. Unpublished notes on File, Centuries Research, Inc., Montrose, Colorado.

2003b Pacomio Chacon: Colorado's Master Sheepherding Artist. Unpublished Paper Read Before the Colorado Council of Professional Archaeologists, Annual Meeting, Durango, Colorado.

Baker, Tess Noel

2004 Canvas in Bark. *The Pagosa Springs Sun*. June 10, 2004.

Blevins, Jason

2008 Aspens Hold Fading Tales of Loneliness. *The Denver Post*, March 3, 2008.

Boner, Jeanette and Kevin Syms (photographer)

2008 Trailing of the Sheep: A World Renowned Festival Just for Ewe(s). *Teton Home and Living*, Fall/Winter, 2008. http://www.lifeinthetetons.com/Teton-Home-and-Living/Fall-Winter-2008/Trailing-of-the-Sheep/ accessed, September 20, 2013.

Born, Phillip L.

1987 Interview with Pacomio M. Chacon, Fruita, Colorado, May 6, 1987. Unpublished notes on file Superintendents Office, Grand Mesa, Uncompahgre, and Gunnison National Forest, Delta, Colorado.

Catholic On Line

2013 http://www.catholic.org/saints/saints.php?sa. Accessed 38/2013.

Chacon, Pacomio and Steven G. Baker

1992 *Pacomio Chacon, Colorado Sheepherder Artist: Master of Nature's Canvases.* Art Show Funded by the Colorado Council for the Arts shown at the Meeker Sheep Dog Trials and Montrose Pavilion, Montrose, Colorado.

Chipeta Chapter, Colorado Archaeological Society

2001 Hot News: Aspen Stories. *The Uncompahgre Journal*, 19(5)4-5.

Colorado Department of Natural Resources

2012 Navajo State Park Hosts Arborglyphs Expert Peggy Bergon on July 14. http://dnr.state.co.us/newsapp/press.asp?pressid=7801. Accessed 3/21/2013.

Curry, Peggy Simson

1951 The Sheepherder in Winter. *The Saturday Evening Post*, January 20, 1951:86.

Douglas, William A. and Richard H. Lane (photographer)

1985 *Basque Sheep Herders of the American West: A Photographic Documentary.* University of Nevada Press, Reno.

Crook, C. H.

1992 Don't Call Him a "Sheepherder." *Country Extra*. May, 1992, 1.

DeKorne, James B.

1970 *Aspen Art in the New Mexico Highlands: A Photo Essay.* Museum of New Mexico Press, Santa Fe.

Finley, Bruce and R. J. Sangosti

2005 Tending Our Flocks: Western Sheep Ranches Depend on Foreign Herders... *The Sunday Denver Post*, June 19, 2005, 1, 10A.

Fishell, Dave

1984 Carving Their Niche. *Jct. Magazine, The Grand Junction Daily Sentinel*, August 5, 1984.

Fletcher, Curley

1986 The Sheep-Herders Lament. In *Songs of the Sage: The Poetry of Curley Fletcher.* Originally Published 1931. Reprinted with a Preface by Hal Cannon. Gibbs M. Smith, Inc., Layton, Utah.

Frost, Robert

1964 *Complete Poems of Robert Frost*. Holt, Rinehart, and Winston. New York et al.

Gilfillan, Archer B.

1929 *Sheep Life on the South Dakota Range*. 1993 Reprint with New Introduction by Richard W. Etulain. Minnesota Historical Society Press, Saint Paul.

Grand Junction Daily Sentinel

1992 Ad for Sheepherder. February 9, 1992.

2009 Obituary for Pacomio M. Chacon, April 14, 1916-July 21, 2009. Friday, July 24, 2009.

Gulliford, Andrew

2007 Reading the Trees: Colorado's Endangered Arborglyps and Aspen Art. *Colorado Heritage*, Autumn, 2007:18-34.

Hammer, Polly

1990 Interview with Pacomio Chacon, Fruita, Colorado, November 16, 1990. Unpublished notes on file Superintendents Office, Grand Mesa, Uncompahgre, and Gunnison National Forest, Delta, Colorado.

Harris, Bill

2010 *Stinking Desert Cairns Project*. MS, Chipeta Chapter, Colorado Archaeological Society, Montrose, Colorado.

2011 The Stone Boys of Delta County. *Montrose Daily Press*, February 18, 2011. Montrose, Colorado.

Hendricks, Rick

2013 Personal e-mail communication from the Office of the New Mexico State Historian in Santa Fe with the author, March 2013.

History Colorado

2015 The Wooden Canvas-Arborglyphs Reflect Hispano Life Along the Pine-Piedra Stock Driveway. http://history Colorado.org/blogs/preservation/2014/10/17/the-wooden-canvas-arborglyphs-...accessed 8/12/2015.

Hunt, Leigh Ann

1994 Interview with Cosme Chacon (brother of Pacomio Chacon), Monticello, Utah, December 7, 1994. Unpublished notes on file Superintendents Office, Grand Mesa, Uncompahgre, and Gunnison National Forest, Delta, Colorado.

2003 Letter of February 28, 2003 from the Forest Archaeologist, Grand Mesa, Uncompahgre, and Gunnison National Forests, Delta, Colorado to Mike Perry, Director, Museum of Western, Colorado, Grand Junction, Colorado concerning the art work of Pacomio Chacon.

Ketelle, Martha, J.

2003 Letter of March 5, 2003 from the Forest Supervisor White River National Forest, Glenwood Springs, Colorado to Mike Perry, Director, Museum of Western, Colorado, Grand Junction, Colorado concerning the art work of Pacomio Chacon.

Lambert, Ruth E.

2014 *The Wooden Canvas: Arborglyphs as Reflections of Hispano Life Along the Pine-Piedra Stock Driveway*. San Juan Mountains Association Press, Durango, Colorado.

Lane, Richard H. and William A. Douglas

1985 *Basque Sheepherders of the American West: A Photographic Documentary*. University of Nevada Press, Reno.

Lucey, Mike

2013 Untitled Sheepherders Poem carved on an aspen tree, Fremont National Forest, Lakeview, Oregon. Published in "Mike Lucey: My Favorite Herder": Arborglyph Recording Project. Posted by Carol Pedersen. http://www.passportintime.com/summaries/97/or97a_sheep.html. accessed 2/24/2013.

Malakoff, David

2013 A History Inscribed on Trees: Tree Carvings Made by Basque Sheepherders Reveal a Little-Known Slice of American History. *American Archaeology* 17(2):19-24.

Mallea-Olaetxe, Joxe

2000 *Speaking Through the Aspens: Basque Tree Carvings in California and Nevada.* University of Nevada Press, Reno and Las Vegas.

Martin, Pat

2005 Tree Carvings: A Living Legacy. *The Fence Post*, Western Slope Edition. January 10, 2005.

Mathers, Michael

1975 *Sheep Herders: Men Alone.* Houghton Mifflin Company, Boston.

McGonagle, Roberta L.

1990 Frank Rodriguez, Sheepherder and Artist. *Women in Natural Resources* 12(1):40-41.

Neal, Shannon Joyce and Dean Humphrey

2001 Alone on the Range. *The Grand Junction Daily Sentinel,* February 18, 2001.

Nix, Steve

2013 Arborglyphs and Tree Carvings: Art and History on Culturally Modified Trees. *About.com Guide.* http://forestry.about.com/od/foresthistory1/a/arborglyph.htm. Accessed 3/22/2013.

O'Conner, Colleen

2005 Outsider Art Comes in From the Cold. *The Denver Post*, Section L, January 23, 2005.

Overturf, Jeff

2003 Letter of March 4, 2003 from the Assistant Regional Archaeologist, Rocky Mountain Region, U.S. Forest Service, Lakewood, Colorado to Mike Perry, Director, Museum of Western, Colorado, Grand Junction, Colorado concerning the art work of Pacomio Chacon.

Parsons, T. S.

1920 Untitled sheepherder's poem. *The Wyoming Stockman-Farmer,* August (1920). Reprinted in Shroup (2008).

Paul, Virginia

1976 *This Was Sheep Ranching Yesterday and Today*. Superior Publishing Company. Seattle.

Pedersen, Carol

1997 "Mike Lucy: my Favorite Herder": Arborglyph Recoding Project.http://www.passportintime.com/summaries/97/or97a_sheep.html. accessed 2/24/2013.

Perry, Michael and Ronna Lee Sharpe

1992 Grant Application to the Colorado Endowment for the Arts from the Museum of Western Colorado, Grand Junction, Colorado on behalf of Pacomio M. Chacon.

Perry, Michael

2003 Grant Application to the State Historical Fund, Colorado Historical Society from the Museum of Western Colorado, Grand Junction, Colorado for the assessment and management of the aspen art panels of Pacomio Chacon.

Peterson, Gwen and Jeane Rhodes

1985 *Cowpunchers, Sheep Herders, and Plain Pig Farmers:WildWest Limericks.* Falcon Press, Helena and Billings, Montana.

Rhodes, Colin

2000 *Outsider Art: Spontaneous Alternatives.* Thames and Hudson, London.

Rumsey, Becky

1992 *Aspen Art Carvings:* Radio Interview with Pacomio Chacon and Steven Baker. High Plains News Service. Interview held at Colorado Public Radio, Grand Junction, Colorado.

Sagstetter, Bill and Beth Sagstetter

2007 The Storybook Trees. *Colorado Heritage*, Autumn, 2007: 30-31.

Schoenian, Susan

2013 Sheep 101. http://www.sheep101.info/sheepbible.html. Accessed 2/28/2013.

Service, Robert

1940 *Collected Poems of Robert Service.* Dodd, Mead & Company, New York.

1987 *Dan McGrew, Sam McGee: The Poems of Robert Service.* Barnes & Noble Books, New York.

Shroup, Allen

2008 Internet Blog, Promised Land: Trailing of the Sheep. http://fourshrops.wordpress .com/2008/11/09/trailing-sheep/. Accessed 4/28/2013.

Siniai, J. Sebastian

1999 Sheepmen a vanishing breed. *The Denver Post,* December 27, 1999.

Stienbeck, John

1939 *Grapes of Wrath.* Viking Press, New York, New York.

Swope, Earle and Amy Nack

2008 Immigrant Shadows: Tracing the Herder's Legacy. Boise, Idaho. http://arborglyph .com. Accessed 4/10/13.

Weidel, Nancy

2001 *Sheepwagon: Home on the Range.* High Plains Press, Glendo, Wyoming.

Wilson Camps

2013 http://WilsonCamps.com/history.html. Accessed March 8,2013.

Wyoming Jobs

2013 http://wyoming.jobs/encampment-wy/sheepherder/33968083/job/. Accessed 2/24/2013.

ABOUT THE AUTHOR

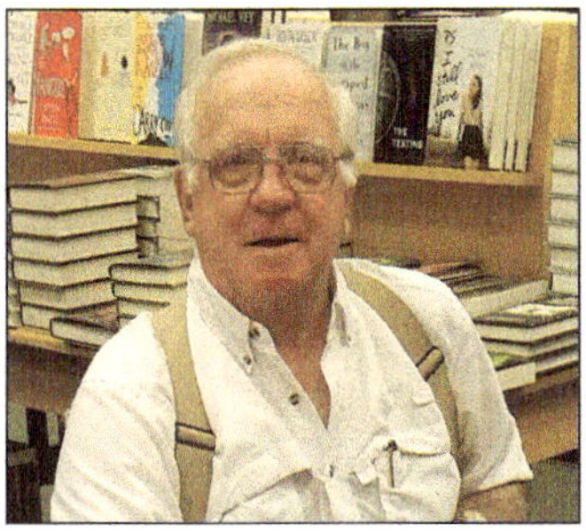

Steven G. Baker B.A., M.A. (b. 1945) is a registered professional archaeologist and ethnohistorian. He has worked full-time in these professions for over fifty years and is now partially retired from Centuries Research, Inc., the consulting firm he founded in 1977. Trained as both an anthropologist and historian he has written hundreds of technical reports, monographs, and journal articles. He has focused much of his career on the local history and archaeology of West-Central Colorado and the region's indigenous Ute Indians. He was the principal author, along with his translator, Rick Hendricks; and illustrator, Gail Carroll Sargent; of *Juan Rivera's Colorado, 1765: The First Spaniards Among the Ute and Paiute Indians on the Trails to Teguayo*... which provides the first descriptions of Colorado's western slope, also published in 2016 by Western Reflections Publishing Company. Baker lives on a small farm near Montrose, Colorado with his wife Nancy Ellen.

CPSIA information can be obtained
at www.ICGtesting.com
Printed in the USA
LVOW05*1222121016
508432LV00007B/8/P